INTERNATIONAL ENERGY POLICY, THE ARCTIC AND THE LAW OF THE SEA

ST. PETERSBURG PAPERS

International Energy Policy, the Arctic and the Law of the Sea is part of a series of publications on oceans law and policy associated with the Center for Oceans Law and Policy, University of Virginia School of Law. This volume is based on presentations made at the Center's Twenty-Eighth Annual Conference co-hosted with the Russian Academy of Liberal Arts Education and the Russian Law of the Sea Association in St. Petersburg, Russia, June 23-26, 2004.

CENTER FOR OCEANS LAW AND POLICY

INTERNATIONAL ENERGY POLICY, THE ARCTIC AND THE LAW OF THE SEA

Edited by

Myron H. Nordquist,

John Norton Moore and

Alexander S. Skaridov

MARTINUS NIJHOFF PUBLISHERS
LEIDEN/BOSTON

A C.I.P. Catalogue record for this book is available from the Library of Congress.

The editors apologize for the inferior quality reproduction of certain graphic materials in this book. We still included these materials since the information remains useful in spite of the imperfect presentation.

Printed on acid-free paper.

ISBN 90 04 14403 X

Koninklijke Brill NV incorporates the imprints Brill Academic Publishers, Martinus Nijhoff Publishers and VSP.

http://www.brill.nl

Printed and bound in The Netherlands.

CONTENTS

Preface

The Twenty-Eighth Annual Conference sponsored by the Center for Oceans Law and Policy, University of Virginia School of Law was held at the Palace of the Grand Duke Vladimir in St. Petersburg, Russia from June 23-26, 2004. The co-sponsors were the Russian Institute for Ocean Law Studies, Russian Academy of Liberal Arts Education, Russian Humanitarian State University and the Russian Law of the Sea Association. The overall focus of the conference was on Russia's ocean interests with particular reference to international energy policy, the Arctic and the law of the sea.

Payel Dzubenko, Deputy Director, Law Department, Russian Ministry of Foreign Affairs, spoke on Russia's continental shelf limits and the navigation regime in the Arctic. He commented upon Russia's recent submission to the Commission on the Outer Limits of the Continental Shelf and emphasized the importance of freedom of navigation in the Arctic. Professor Kuen-chen Fu of Xiamen University Law School addressed himself to China's growing demand for energy resources, much of which must be supplied from rich offshore oil and gas fields in the region. Among other matters, Dr. Fu reviewed China's 1982 regulations that govern foreign enterprises doing business with the China National Offshore Oil Corporation. Boris A. Smuslov, who holds the Maritime Law Chair, Maritime State Technical University in St. Petersburg, identified many inadequacies in the global fight against terrorism at sea. The Director of the Aberdeen Institute of Coastal Science and Management, William Ritchie, explained the implications of the environmental policy of the "transfer of good practice" in major offshore oil and gas developments. Russia's environmental regulations pertaining to offshore oil exploration were detailed by Richard N. Dean and Michael P. Barry, lawyer and economist respectively for the international law firm, Coudert Brothers LLP. The co-authors reviewed the incentives for Russia to further develop its vast energy resources, which must be balanced with the potential harm to the environment. They cited risks posed by tanker accidents, discharge of drilling muds and cuttings, pipeline spills, high energy seismic surveys, and especially sensitive environments. Lastly, Dean and Barry offered three approaches to environmental regulation: public law, multilateral institutions and private contractual arrangements. Professor Paul B. Stephan of the University of Virginia School of Law spoke on the challenges law faces in dealing with the large capital investment and huge economic risk that accompanies energy development and distribution. He selected discrete legal issues to illustrate ways in which transactional lawyers could assist in strengthening the legal protections for their clients while advising on the limits of such protections.

The Black Sea and the straits connecting it to the open ocean play a significant role in the transportation of energy resources from the region, including the states bordering the Caspian Sea. Vladimir S. Kotliar of Moscow's Diplomatic Academy criticized Turkish policy in taking what he characterized as unilateral actions concerning commercial navigation through the Straits. Dr. Nilüfer Oral, Assistant Director of the Istanbul Bilgi Marine Law and Policy Research Center, countered with a defense of Turkish policy outlining the steps taken in the International Maritime Organization (IMO) and in the region to enhance safety of navigation and the regulation of maritime traffic. Cooperation between Vessel Traffic Services in the Black Sea was explained by Captains Robert Hofstee and Ozkan Poyraz. They concluded with an appeal for even greater coordination between port States, flag States, classification societies, and other entities charged with the safety of shipping in the Black Sea area. An up-to-date report on oil spill contingency planning and technical cooperation in this area was next given by John Østergaard, Senior Advisor on Marine Pollution, Marine Environmental Division, IMO. With IMO's technical support, all Black Sea States have or are finalizing a modern national oil spill contingency plan based on internationally adopted standards.

The unique challenges in the Arctic region were highlighted by Ambassador Sverrir Haukur Gunnlaugsson, Chairman of the Board of the Law of the Sea Institute of Iceland. Dr. Larry Mayer and several colleagues at the University of New Hampshire's Center for Coastal and Ocean Mapping looked at collecting data to support potential claims in the Arctic under Article 76 of the 1982 Convention on the Law of the Sea (UNCLOS) for Canada, Denmark, Norway, Russia and the United States. The authors concluded that despite the progress in the compilation of gravity, magnetics, seismic and bathymetric data in the Arctic, compilations were still inadequate to support a claim under UNCLOS. Alexander S. and Mariya A. Skaridov surveyed the legal aspects of Russia's Arctic continental shelf claims. Several issues raised in Russia's submission to the Commission on the Limits of the Continental Shelf were noted, including the high potential for hydrocarbon deposits in the area likely to be claimed by Russia in the Arctic. Alexander Klepikov of the Ocean and Atmosphere Department, Arctic and Antarctic Research Institute, St. Petersburg and two colleagues covered the research studies on rapid Arctic climate change undertaken as part of the Arctic Monitoring and Assessment Program. An Icelandic perspective on climate change and a northern sea route was given by Thor Edward Jakobsson of the Icelandic Meteorological Office. Dr. Jakobsson described the possibility of a trans-Arctic sea route and the implications for the future. Franklyn Griffiths, an emeritus professor of political science at the University of Toronto, argued that the Northwest Passage was not a live option for intercontinental navigation. He proposed several areas where the United States and Canada should, as a practical matter, cooperate much more than they have to date but was skeptical of those with an "overwrought view of climate-change effects on archipelagic sea ice." The Chair of the U.S. Arctic Research Commission, George B. Newton, examined several plans for Arctic transportation systems under existing and possible future developments. He urged greater attention to oil spill response methodologies based on research and international cooperation.

The third cluster of themes at the St. Petersburg Conference dealt with the current law of the sea and the status of UNCLOS. No one is better qualified to speak on the current and emerging issues in the law of the sea than Satya N. Nandan, Secretary-General of the International Seabed Authority. In his keynote remarks to the Gala Dinner audience, Ambassador Nandan noted that 2004 marked the 10th Anniversary of the entry into force of UNCLOS and that 145 States were party to the Convention. He mentioned the importance of the 1994 Part XI Agreement and the 1996 Fish Stocks Agreement, both of which were negotiated under his leadership. Nandan was critical of the use of flags of convenience to escape responsibilities under the Convention, especially in the field of fisheries. He pointed out that there are emerging threats to biodiversity in the deep ocean and no agreement in the international community on how to handle the problems posed by human activities. Lastly, Nandan flagged the increasingly serious issues of piracy and armed robbery at sea and the related question of the prevention and suppression of terrorism against shipping. The perspectives of the United States were provided by Margaret F. Hayes, who directs the Office of Ocean Affairs, U.S. Department of State. She reviewed major global and regional developments in ocean policy, with particular attention to the Arctic region. The prospects for U.S. accession to the Convention were covered by Michael J. Mattler, Deputy Chief Counsel, U.S. Senate Committee on Foreign Relations. He reported that the efforts in the U.S. Senate during the past year have placed the United States closer to joining the Convention than previously. Professor Aldo Chircop, presently holding the Canadian Chair in Marine Environment Protection, World Maritime University, Malmo, Sweden also spoke on a leading edge issue in the law of the sea, namely Article 82 of UNCLOS. This article lays out the manner in which coastal States are to make payments to the International Seabed Authority for the exploitation of non-living resources on the continental shelf beyond the 200-mile zone.

The St. Petersburg meeting closed with remarks by Professor John Norton Moore, who highlighted the great importance of the rule of law in the oceans and the contribution to that end made at this conference.

Note by the Editors

The Conference on International Energy Policy, the Arctic and the Law of the Sea naturally focused on Russia since the meeting was held in the beautiful and historic international port of St. Petersburg. Local arrangements were almost entirely personally made by Professor Skaridov, who heads the Russian Academy of Liberal Arts Education and directs its Institute for Ocean Law Studies. Dr. Skaridov was given invaluable help from Judge Anatoliy L. Kolodkin, who chairs the Russian International Law Association as well as the Russian Association of the Law of the Sea. Their combined efforts resulted in attracting panelists and participants with first-class expertise, especially from within Russia. Not all speakers were able to submit papers in English text and we regretfully note that the proceedings capture most but not all of the content actually presented at the Conference.

Judge Kolodkin and Professor Skaridov were assisted in St. Petersburg by Alexander Etimal, Glendamere Ivanov, Tatiyana Mordvinova, Anatoli Butkoca and Nabila Notiava. Alina Ilyina deserves to be singled out for her brilliant simultaneous translations that garnered universal accolades. The conference site was impressive taking place as it did in the palace of the Grand Duke Vladimir in the heart of the largely preserved 18th century buildings and gracious canals of St. Petersburg. No one who attended the meeting will ever forget the Russian artistic heritage and cultured atmosphere we found. St. Petersburg is truly one of the great cities in the world.

The Virginia team is led by Professor John Norton Moore, who directs the Center for Oceans Law and Policy, and by Professor Myron H. Nordquist, the Center's Associate Director and Editor. The real work on administrative arrangements, however, was done by Donna D. Ganoe, who is the Center's Executive Administrator, and her assistants, Kay W. Wood and Kathy H. Wood. While responsibility for the final text in this volume rests with the three co-editors, Myron H. Nordquist, John Norton Moore and Alexander S. Skaridov, Elyse H. Hunter of the Center deserves particular recognition for her conscientious technical editing that was indispensable for a camera-ready finished product.

Most of the funds to sponsor the Conference were provided either by the Russian hosts or by the Center for Oceans Law and Policy. Additional support was given, however, by the Law of the Sea Institute of Iceland, the World Maritime University in Malmo and Istanbul Bilgi University.

Our collective thanks to all sponsors, hosts and individuals who helped make the Twenty-Eighth Annual Conference a memorable success. We hope that our modest efforts will foster greater understanding between Russia and the West about international energy policy, the Arctic and the law of the sea and thereby contribute to rule of law in the world's oceans, whether liquid or frozen.

THE EDITORS

INTRODUCTION AND WELCOMING REMARKS

Introduction and Welcoming Remarks

John Norton Moore[*]

Welcome Deputy Director Dzubenko, Secretary-General Nandan, Judge Kolodkin, Ambassador Larsen, Professor Skaridov, and distinguished participants and guests. I am John Norton Moore of the Center for Oceans Law and Policy of the University of Virginia and I would like to welcome all of you to this conference on International Energy Policy, the Arctic and the Law of the Sea. A special thanks to the co-sponsors of this conference, the Russian Academy of Liberal Arts Education and the Russian Law of the Sea Association. On a more personal level, we owe special thanks to our Russian hosts. We especially thank Judge Anatoliy L. Kolodkin, a Judge of the International Tribunal on the Law of the Sea, President of the Maritime Law Association and the Russian International Law Association, Chairman of the Russian Law of the Sea Association, Deputy Director General of State Research Institute of Marine Transport, and a Member of the Scientific Council to the Minister of International Affairs of the Russian Federation. Today the world acknowledges Judge Anatoliy Kolodkin as the best-known international lawyer in Russia.

Secondly, I would like to thank Captain Alexander S. Skaridov, formerly head of the legal department of the Russian Naval Academy and current head of the Academy of Liberal Arts of Communications Technologies Education, St. Petersburg, of the Russian Academy of Humanitarian Sciences. It is Professor Skaridov who has done most of the wonderful work that has produced this conference, along with Myron Nordquist of the Center for Oceans Law and Policy.

It is a special pleasure to hold this conference in Russia and particularly in St. Petersburg. The world has long admired the rich cultural and artistic heritage of the people of Russia, a heritage including writers, poets and composers, such as Tolstoy, Chekhov, Pushkin, Dostoevsky, Tchaikovsky and Bronski and, of course, the artistic brilliance of a Bolshoi Ballet. In St. Petersburg, one of the great cities of the world, it is especially appropriate to recall the great contribution of Alexander

[*] Professor of Law and Director, Center for Oceans Law and Policy, University of Virginia School of Law.

Sergeevich Pushkin, who died here in 1837. His magnificent works stand as the cornerstone of Russian literature or, as Gorsky put it, "the beginning of beginnings."

It is also altogether appropriate that United States and Russian institutions jointly collaborated on this conference. Even during the height of the Cold War, we worked together on the Law of the Sea. And we have long shared an interest in the protection of navigational freedom, among other oceans issues. Indeed, in common we have understood that the community interest in navigational freedom is the most important common heritage of mankind in the ocean. I had the privilege of working from the U.S. side in our Law of the Sea collaboration. But for that robust collaboration, which sometimes surprised our colleagues in Washington and Moscow, there would likely never have been a third United Nations Conference and successful treaty on the Law of the Sea.

More recently, I am especially pleased to say that our nations have shared a common interest in democracy and the rule of law. Indeed, in 1990 it was my privilege again to represent the U.S. side in our first joint talks on the rule of law in Moscow, and in this beautiful city, St. Petersburg.

The subject of our conference is a dual focus on issues relating to the Law of the Sea in both parts of that focus: international energy policy and the Arctic. The importance of international energy policy hardly needs discussion. Oil is the number one commodity in world trade and is essentially linked to economic development and economic growth around the world. Enhancing stable supply is of great importance to every nation on earth. Indeed, perhaps nothing is more important economically to our joint economic development than stable supplies of oil and energy for the world. We have seen what has happened in the oil shocks in the 1970s, first the doubling of oil prices in 1973-74 followed with a doubling of oil prices again in 1978-79. Once again we saw a huge spike in oil prices associated with the Gulf War in 1990 and, to a lesser extent, we are witnessing an oil spike associated with events in the Middle East in 2003 and 2004. The doubling of the oil price in the two oil shocks of the 1970s had a particularly devastating effect on the global economy. If you look at India and the second oil shock of 1978-79, India had a greater outflow of wealth in that one year of the second oil shock than it did in the entire period of colonialism. And many of the developing countries of the world went into severe debt that still lingers. The question of international energy policy is of enormous importance and the importance of Russia to that subject is without question. Russia has perhaps the greatest reserves of oil in the world and is the largest producer of oil in the world.

Let me shift to the Arctic issues. There are many different components of this: navigation, technological challenges of working in ice-covered areas, either in relation to navigation and maritime interests or in relation to oil and natural gas platforms (for example, issues of the environment), issues of science, and many others, though Arctic issues are not the immediate concerns of people around the world as much as international energy policy. However, the Arctic is very important, and the great importance of Russia in relation to the Arctic is evident. Russia has, with the Northern Sea Route, the greatest interest in navigation in ice-covered areas of any nation in the world. It is the only nation in the world that has developed nuclear-powered icebreakers. Since the United States also shares an Arctic boundary, we have a shared interest in things Arctic.

We have a world-class group of experts to address these issues of international energy policy and Arctic matters related to the Law of the Sea, and I am confident that this conference will make a contribution to the understanding of both issues. At this point it is my privilege to call on the honorable Judge Anatoliy Kolodkin, to welcome you on behalf of the Russian Association of the Law of the Sea and our Russian Hosts.

Introduction and Welcoming Remarks

H.E. Anatoliy L. Kolodkin[*]

First, let me congratulate Professor John Norton Moore, Professor Myron H. Nordquist and my colleagues on this international law conference organized by the Center for Oceans Law and Policy of the University of Virginia School of Law and the St. Petersburg branch of the Russian Association of the Law of the Sea.

Secondly I want to express our deepest gratitude to Professors Moore and Nordquist for the Center's leadership in suggesting the holding of this conference in St. Petersburg, Russia. I would also like to commend Professor Alexander S. Skaridov, Chairman of our Association's St. Petersburg branch, who carried out most of the logistic and organizational matters.

One cannot dismiss the importance of holding the conference in St. Petersburg, a city where there has been a focus on international law, in particular on the Law of the Sea. In this city, most Russian institutions are involved in Law of the Sea studies, including the Law Faculty of the State University, the Admiral Makarov Academy, the Naval Academy, and the Institute for Ocean Law Studies of the Russian Academy for Humanitarian Education. Research projects are strongly related with the needs of industry and business, including the St. Petersburg sea port, which is in the process of a capital reconstruction project. That sea port is why St. Petersburg became the sea capital of Russia. In this city, great historical thoughts on general international law were born.

Thanks to Russia's initiative, the first Peace Conference took place in the Hague in 1899. In 100 years Russian initiatives have led to the holding of three conferences; the first took place in the Hague, the second in St. Petersburg, organized with Netherlands colleagues by our Associations of International and Sea Law; and the third today.

The subject of our conference has been chosen at a good time. Though the UN Convention on the Law of the Sea includes 145 participants, including Canada, there are some negative themes.

---

[*] Judge, ITLOS, Chairman of the Russian International Law Association, Chairman, Russian Association of the Law of the Sea.

In his speech in Kazan, Russian President Vladimir Putin stressed that one must follow international law standards in developing an up-to-date international relationship.

Yet we have observed some countries acting in such a way that they break the universal contract, namely the UN Convention on the Law of the Sea.

The unlawful practice in limiting tanker shipping in 200 mile exclusive economic zones testifies of that, and it is clear by the topic of our conference that notification of the passage through these zones during 48 hours, ignoring the points of Convention, is what leads to the UN International Tribunal's solutions in the Law of the Sea for urgent discharge of arrested ship and crews.

The topic of the conference is a matter of current interest in light of the conservation regime for non-proliferation of nuclear weapons, unlawful sea transportation of this kind, the struggle against terrorism and piracy on the sea.

It is of great importance that the participants of this conference are prominent scientists, government representatives, foreign policy makers, and high-ranking officials from the United States, Russia, Great Britain, Canada, Netherlands, Norway, Sweden, Ireland, Iceland, Turkey, Latvia and other countries.

Honorable conference participants, let me promise you that our conference will be of service to the international cooperation in constructing regulations for operating the sea resources of energy and to the well-being of our people.

KEYNOTE ADDRESS

Current and Emerging Issues in the Law of the Sea

Keynote Speaker
H.E. Satya N. Nandan[*]

First let me thank Captain Alexander S. Skaridov for hosting this meeting and making it a very successful event. As always, we should all be grateful to Professor John Norton Moore for convening this conference under the auspices of the Center for Oceans Law and Policy of the University of Virginia. This is probably the most important annual conference on the Law of the Sea that takes place in different centers around the world. It helps us to promote the Convention, it brings together experts from different parts of the world, which provides an opportunity for overseas and local experts to interact. The Center makes a very important contribution in support of the Law of the Sea Convention and the rule of law in the oceans. I also want to thank Judge Anatoliy Kolodkin, whom I have known for a long time and who I know is keenly interested in the Law of the Sea. I am also very pleased to meet two of my friends in Russia, Vladimir Kotliar and Payel Dzubenko.

This year marks the 10th anniversary of the entry into force of the 1982 UN Convention on the Law of the Sea. The anniversary actually falls on November 16, 2004. As you well know, the Convention now has 145 parties to it from a possible 195 states. It is universally applied by both parties and non-parties. In 1970, when we set out in the Seabed Committee to prepare for the Third United Nations Conference on the Law of the Sea, none of us conceived in our wildest dreams that we would have such a successful Convention. At that time, ocean law was in a chaos. Today there is widespread uniformity in State Practice. For instance, we have 140 states with territorial seas of up to 12 nautical miles and 117 states have declared a 200 mile exclusive economic zone. There are about 12 states whose territorial sea claims do not conform to the Convention, but most of them are trying to make necessary changes in their national legislation to allow them to become parties. During this conference one of the most important statements for me was that made by Michael Mattler, when he indicated how close the United States is to becoming a party to the Convention. That would be a significant step in

[*] Secretary-General, International Seabed Authority.

the development of the Convention and its broad acceptance and application. I have already told Mike that he is doing fantastic work as a staffer, and I hope that he will convey our appreciation to his colleague, Ken Meyers who is chief of staff of the Senate Foreign Relations Committee under U.S. Senator Richard G. Lugar.

The process of negotiating the Law of the Sea Convention was painstaking. We tried to achieve a consensus on every major substantive issue. By 1982 it was very clear that we had, for the most part, achieved our goal. But Part XI, the Deep Seabed Mining part of the Convention, remained controversial. When you have a legal vacuum such as we had in respect of the mineral resources in the deep seabed, it provides a rich ground for legal, political, and ideological differences. It did not surprise me, therefore, that there was a need for some time off so that the passions could cool while we found reasonable, practical, and realistic solutions. We were able to achieve that in 1994 through the Part XI Implementation Agreement. That agreement was adopted by consensus and was signed by all states that participated in the negotiations, including the United States. That agreement has become the basis for the regime for deep seabed mining and by its terms it is an integral part of the Convention.

As we proceed in the implementation of the Convention, one major challenge for states is to protect the integrity of the Convention. Its strength lies in the balance it has achieved with respect to competing national and international interests in the oceans and in the competing uses of the oceans. The challenge, therefore, is to insure that the Convention is applied by states in a manner consistent with its provisions. Too often states have readily asserted their rights to jurisdiction over maritime areas without regard to the corresponding duties that are attached to those rights. This trend is becoming increasingly apparent in a number of areas, and is giving rise to serious issues at national and international levels. As a consequence, a number of new issues have to be addressed within the framework of the Convention. In the past, we have dealt with at least two major issues that needed to be resolved through "implementation agreements." The first was the 1994 Part XI Agreement, and the second was the 1996 Fish Stocks Agreement. Other problems that are emerging should also be addressed in the same manner and in a forum under the auspices of the United Nations rather than in organizations that have only sectoral concerns and competence so that the overall balance achieved in the Convention is not compromised.

Some current issues that are being discussed in various Law of the Sea forums: The first issue is the failure in fisheries governance. The continuing decline in fish stocks implies that the current system in fisheries management, both in national areas and in the high seas, is not working. This dysfunction is due to a lack of acceptance of the duties under the Convention and an ineffective discharge of the implementation responsibilities at the national level and through regional fisheries management organizations. A remedy lies in the implementation of sensible management principles contained in the FAO Code of Conduct for Fisheries Management and in the 1996 UN Fish Stocks Agreement. These principles are a better basis for conservation and management of fish stocks, and include an ecosystem approach as well as the use of a precautionary approach to fisheries management.

Many important fish stocks in national areas have been grossly mismanaged by coastal States who through defective management structures, subsidies and a lack of political will have allowed their EEZs to be over fished. The high seas areas have also been subjected to over fishing by distant water fishing states. States have failed to exercise control over their vessels. Flag State jurisdiction can only work if the flag State has the ability to effectively control vessels flying its flags. This is key in the discharge of flag State responsibilities under many provisions of the Convention. Where the link between a vessel and its purported or ostensible flag State is tenuous, or, in effect, nonexistent, it is apparent that the use of such flag States by owners of vessels is no more than a device to escape the responsibilities under the Convention or from conservation measures, such as quotas, adopted by regional fisheries organizations. They also seek to avoid the responsibilities that arise from the labor and safety standards. The time has come for the international community to meet this problem head on. It is time to establish a principle of the co-responsibility of the flag State and of the nationality of the owners and operators of fishing vessels. This is not a novel idea. Precedent exists in cases of oil pollution where owners of vessels are equally liable with flag States for damage caused by their tankers. The problem in the fisheries industry is acute. Five out of seven cases relating to fisheries that have come to the International Tribunal for the Law of the Sea concern violations of fisheries regulations by fishing vessels flying flags of convenience. This does not take into account the hundreds or perhaps thousands of cases that do not come before any tribunal. Illegal, unregulated and unreported fishing remains a major obstacle to effective fisheries management. It has been a matter of serious concern at FAO and at the United Nations General Assembly.

The second item currently being discussed in different fora is the conservation and management of the biodiversity of the seabed beyond national jurisdiction. There is increasing awareness of the rich biological diversity in the deep ocean. There is concern about the threats posed by human activities in relation to biological diversity in the deep ocean. The plan of implementation of the Johannesburg World Summit on Sustainable Development recommended that states should take action to maintain the productivity and biodiversity of important marine and coastal areas, including areas within and beyond national jurisdiction through the use of an ecosystem approach to ocean management, elimination of destructive fishing practices, and the establishment of marine protected areas. The UN General Assembly endorsed the Summit's implementation plan and urged that there was pressing need to address the management risks to the marine biodiversity of seamounts, cold water coral reefs, and certain other underwater features. The issue is also being dealt with in the context of the Convention on Biological Diversity and by the International Seabed Authority in its regulations for the exploration of minerals on hydrothermal vents and cobalt crusts on seamounts. As far as marine protected areas are concerned, a number of states have established such areas in their national jurisdictions. Examples of these are Australia, Canada, New Zealand and Norway. It is estimated that there are some 10,000 to 30,000 seamounts worldwide, most of which are in the international area. They are considered biological hotspots with high species diversity and endowment. Seamounts are under increasing pressure from deep sea fishing and they have been impacted by physical damage from trawl fishing. There are examples of seamounts that had been very rich in biodiversity, with corals and sponges and other types of biological species, which were rendered bare, like moon surface, of all fauna and flora by the use of deep sea trawl nets. Where this has occurred it has not only destroyed the biodiversity of the seamount, but also destroyed the habitat of valuable fish stock such as the orange roughy. It takes orange roughy about 100 years to mature and regenerate. In response to this tragedy in the oceans there is a move to ban all bottom trawling on seamounts.

A third issue is the conservation of the biodiversity that exists around hydrothermal vents. The vent ecosystem includes micro and macro organisms such as giant tube worms, clams, shrimp, crabs, and mussels clustering around the vents. In addition to the environmental aspect of this ecosystem in the context of mineral development, a new issue has arisen as a result of bio-prospecting. It has become increasingly common for marine scientific research activities, especially those related to biological and geological sampling, to have links to onshore commercial activities. The intensification of research into commercially useful genetic resources is raising significant legal and institutional issues. There is an apparent gap in the Convention with respect to the use of genetic resources of the deep seabed for commercial purposes. The discussions concerning this gap are taking place in a number of different forums. Because of the exploitative purpose and profit-making goals, many states compare bio-prospecting with prospecting for minerals. Article 143 of the Convention provides that marine scientific research in the deep seabed area shall be carried out for the benefit of mankind as a whole. The question that is raised is how mankind can benefit from the commercial development of these resources. A parallel is being drawn with the regime for minerals in the deep seabed, which has been declared the common heritage of mankind.

There are a number of other current that I would like to mention briefly. One is piracy and armed robbery at sea. The indications are that the problem is becoming very acute and is interfering with safe navigation. Piracy incidents grew to 445 in 2003 from 370 in 2002. Some 31 seafarers were killed and about 80 injured. The issue is a very live one, and it is creating new financial and administrative burdens, especially on strait States and coastal States in congested traffic areas.

A second issue is the prevention and suppression of terrorism against shipping. This is being considered by the IMO, and amendments are being made to the annexes of the Safety of Navigation Agreement. One important development is that the original agreement requires continuous synopsis or records to be provided to the flag States, such as records of ship registration. Another important issue that we will hear more about in the future is the trafficking in weapons of mass destruction. This is a new development that pertains to the interdiction of suspect vessels on the high seas. In 2003 some 11 states initiated a program on dealing with this issue. This was aired in the United Nations Security Council with a declaration by the heads of States when they met in 1992. That declaration stated that proliferation of all weapons of mass destruction constitutes a threat to international peace and security and underlines the need to prevent transportation of such weapons.

Let me conclude with my earlier theme that the Convention has enjoyed a remarkable success in 10 years since its entry into force. There is, broadly speaking, considerable consistency in State Practice. We must continue to ensure that the Convention is applied uniformly. The Convention has built-in flexibility, and new issues that arise can be resolved within its framework. We have already adopted two implementation agreements without disturbing its basic structure and balance. New issues should be dealt with in a similar way. If we follow this approach, the Convention will continue to serve mankind by promoting good order in the oceans and by contributing to international peace and security.

PANEL I:
GLOBAL PERSPECTIVES

Russian Maritime Policy and Energy Resources

Payel Dzubenko[*]

Russian marine policy in the Arctic is strongly related with two basic issues: Russian continental shelf limits and the regime of Arctic navigation.

Russia made a submission to the Commission on the Limits of the Continental Shelf. At the time this paper was written, the Commission had finished its first consideration of Russia's application on matters connected with borders of the continental shelf. The Commission paid close attention to data and Russia's views, and a number of questions arose concerning side limits of the continental shelf and grounds of the central part of Russia's continental shelves bordering 200 miles in the Arctic Ocean and in the Arctic as a whole.

Russia was asked to present more information, and we are working to be ready to give final answers. We will then be able to make final decisions and juridical reform of our shelf's border.

Lomonosov and Mendeleyev are in a large geologic-geophysical region, and as such there are some juridical consequences. As a result of UNCLOS, if those are just ocean's raisings, i.e., ocean's ridges, then it will not be difficult to overcome obstacles because they are continental raisings, and the ocean spreads to their sides. Natural scientific forums took place, including one in St. Petersburg, and everyone agreed that it is possible to expand our continental relation toward the North Pole.

Russia has no ambitions on the continental shelf in the South. There is not a shelf within 200 miles on the Baltic Sea or the Black Sea.

In the East, Russia has to settle some matters. Relations between Russia and Japan are not yet regulated. These issues are not of a legal nature, but regard the method of marking off our continental shelf and differing opinions have nothing to do with our application.

The land bordering the United States on the Bering and Okhotsk Seas (there are also some parts of shelf at a distance of more then 200 miles) is not regulated, either. To the north, from the Bering Strait, which goes toward the meridian of the North Pole, it is quite settled. There are no

[*] Deputy Director, Law Department, Russian Ministry of Foreign Affairs.

obstacles, but not because of a 1990 agreement about the bordering since the agreement has not yet been ratified. A line about Alaska's seals appeared in 1867 in the Convention, perhaps without any geographical description but very clearly marked, and was further used in the 1970s for fish protection measures. In this region fishermen are not satisfied, because Russia has lost most traditional fishing locations.

The situation in the West is much worse. Contrasting with the East, where there are currently no oil or gas structures at all, this region is rich with mineral resources (Khodinsk, et al.). We have many serious issues to address, for example in negotiations with Norway, about places in the Barents Sea. For a long period of time we have had a moratorium on delimitation of the whole disputable region (the central part of the Barents Sea) for the development of mineral recourses. The origin of the moratorium evokes great interest, of course. Why and how did it come about? Can someone begin prospecting work and collaboration there? In the beginning, we decided to operate only with the available data and not to work on obtaining new data. We hope the negotiations are becoming active for the best perspectives. At that point we will be able to cooperate with Norway in this sphere, because these deposits on the bordering line are to be compared with the Caspian model, only somewhat different.

Another question that seems to bring up a very important point deals with sea transportation of energy power and other resources. It concerns sea navigation as well. The principle of sea freedom has been legalized outside the territorial waters, including in the economic zones of countries. Yet it has already become a tradition to make burdensome sea navigation regimes in the economic zones, and in this connection one should note the details of working out the Convention, because different countries had different approaches.

The Convention succeeded in uniting all interests, and it is effective now. During the Conference, the 200-mile exclusive economic zones were acknowledged by the Soviet Union. But the economic zone remains an open sea part, especially concerning shipping.

We are not against defense of the sea's surroundings, yet the measures must be reasonable and rationally motivated. Some countries unilaterally took such measures. They have created specially guarded regions, and we welcome this initiative, if later they act in accordance with competent international organizations of universal character. In fact, it is very difficult for a captain to get his bearings after every 100 miles; this would be a different kind of navigation.

As for the protection of the marine environment, we support the idea of making protected sea areas near the Galapagos Island and the Great Barrier Reef. However, we are worried because vast areas of the world ocean are declared sea reserves unnecessarily, e.g., the Baltic Sea. This causes great problems for sea trade. We have information that indicates there are some similar projects concerning the Black Sea, the Mediterranean Sea and a major portion of the Western Atlantic. We believe that our anxiety is quite appropriate because these areas of the world ocean are very important for trade.

I suppose all measures taken to protect the world ocean should be aimed at saving nature, but not at promoting commercial interests only for some countries. These measures should not harm free trade between countries.

Current Oceans Policy: United States Perspectives

Margaret F. Hayes[*]

INTRODUCTION

This paper will review current oceans policy first at the global level; then at the regional level, with particular attention to the Arctic region; and then at the national level, tracking recent developments within the United States.

GLOBAL OCEANS POLICY

November 2004 marks the 10th anniversary of the coming into force of the United Nations Convention on the Law of the Sea, the overarching framework within which virtually all policies relating to the oceans are developed. The parties to the Convention total 145, a number that supports a claim of near-universality.[1] This paper will discuss prospects for increasing that number before the anniversary date.

During the past 10 years the three institutions created under the Convention have found their homes, set up their procedures, and begun to carry out the functions assigned to them. The International Tribunal on the Law of the Sea has dealt with almost a dozen cases in an efficient and restrained manner. Under the skilled leadership of Ambassador Satya N. Nandan, the International Seabed Authority finalized plans of work for the seven pioneer investors. The ISBA also adopted regulations on prospecting and exploration of polymetallic nodules, and organized workshops to collect data and information, which are stored in its database. The Commission on the Limits of the Continental Shelf has reviewed the first submission, from the Russian Federation, received a submission from Brazil that will be considered at the end of summer 2004, and will have at least one more submission, from Australia, by the Convention's anniversary date.

---

[*] Director, Office of Oceans Affairs, U.S. Department of State.

The most important global institution for promoting maritime safety and protection of the marine environment, the International Maritime Organization, has addressed recent terrorist threats by adopting a new maritime security regime that is on the verge of entering into force,[2] and by working on a draft protocol to the Convention for the Suppression of Unlawful Acts against the Safety of Maritime Navigation. The IMO has also accelerated its phase-out schedule for single-hull tankers, reviewed proposals for designating Particularly Sensitive Sea Areas, and adopted a convention for control of ships' ballast water and sediments. Realizing that international agreements are of little value unless they are implemented and enforced, IMO members are developing an Audit Scheme to gauge the effectiveness of flag States, port States, and coastal States in carrying out their obligations under IMO instruments.

Another indicator of the maturing of oceans policy is the coming into force of the 1995 United Nations Fish Stocks Agreement, along with the revisions that are being made to regional fisheries management agreements to incorporate its principles. Yet overfishing and habitat destruction continue to threaten the long-term health of the world's fish stocks. States that license vessels to fish within their own exclusive economic zones or on the high seas must be held accountable when those vessels engage in illegal, unregulated, or unreported (IUU) fishing. Implementation of the Food and Agriculture Organization's international plans of action relating to fisheries will be a big step in the right direction.

The United Nations General Assembly's deliberations on oceans have also matured, through the evolution of the Informal Consultative Process held each spring for the past five years. This week-long session provides a forum for delegates to learn about new developments in the oceans world and to discuss possible approaches to maritime issues. The report of this meeting now forms the basis of the annual UNGA resolution on oceans and law of the sea, which has become a less repetitive and more forward-looking and useful document in the last few years.

The Informal Consultative Process meeting that concluded in June was to focus on protection of high seas biodiversity and on new uses of the oceans for energy production. High seas biodiversity attracted much more attention than new energy sources, as environmental groups called for the participants to recommend that the General Assembly establish a moratorium on bottom-trawling in areas beyond national jurisdiction, and a large number of delegations proposed that deep-sea genetic resources should be considered part of the "common heritage of mankind" under

Part XI of the Convention. This interpretation, which has little foundation in the text or negotiating history of the Convention, would equate living organisms found at hydrothermal vents or around other seabed features with the carefully defined mineral resources to which the Part XI regime applies. The United States and other delegations vigorously refuted this proposition, but the proponents persisted in trying to raise the subject at the States-Parties meeting the following week.

A day and a half during the Informal Consultative Process was allocated to an international workshop on the Global Marine Assessment. The workshop reviewed a draft document on a process for global reporting and assessment of the marine environment, including peer review, a secretariat, capacity-building, and funding. Unfortunately, a dispute over the scope of the Assessment – whether it should include living marine resources – is standing in the way of launching this initiative.

REGIONAL OCEANS POLICY

Turning from the global to the regional level, this paper has already mentioned the work of Regional Fishery Management Organizations, which are promoting the sustainability of fisheries through the use of increased knowledge of fish stocks and their interactions with the marine environment, through ecosystem management and precautionary approaches, and through creative means of enlisting port States and importing nations in the effort to combat IUU fishing.

Regional approaches are also critical in protecting the marine environment. UNEP's regional seas programs are active, to varying degrees, in increasing the capacity of coastal States to combat the major causes of environmental degradation. Due to our special ties to two regions, the United States participates most actively in the Pacific and the Caribbean programs. The Caribbean Environment Programme recently joined with the United States, and with many other governments, international organizations, universities, financial institutions, non-governmental organizations, and corporations, to form the White Water to Blue Water Initiative.

White Water to Blue Water aims to stimulate partnerships that will advance integrated watershed and marine ecosystem-based management. It promotes regional cooperation and strengthens developing country capacity:

- to address land-based sources of marine pollution (e.g., sewage and industrial discharges, agricultural run-off);
- to build sustainable fisheries;
- to improve agricultural and forestry practices;
- to meet challenges associated with tourism; and
- to prevent degradation of coastal areas.

White Water to Blue Water, launched at the World Summit on Sustainable Development, first focused on the Wider Caribbean region. A Partnership Conference was held in March 2004 in Miami, with more than 700 people from 32 countries attending. More than 100 partnerships have been formed in the past two years in connection with this initiative. We believe the White Water to Blue Water outcomes in the Caribbean may serve as a blueprint for future efforts in Africa, in the South Pacific, and elsewhere.[3]

Two initiatives in the Arctic deal with energy and with ocean policy. Both initiatives are activities of the Arctic Council, the intergovernmental forum consisting of eight member States, each with Arctic territory, as well as representatives of people indigenous to the Arctic. The United States is fully engaged in both activities. I bring these to your attention because they illustrate the ways in which voluntary cooperation among like-minded states can contribute to good governance in the Arctic.

The first initiative is the "Assessment of Potential Impacts of Oil and Gas Activities in the Arctic." The Assessment is being prepared in response to a request from the Ministers and Senior Arctic Officials of the Arctic Council. They have asked for a report, to be prepared in time for the Ministerial meeting in 2006, that builds on and expands an earlier assessment completed in 1997. The United States and Norway have agreed to co-chair the Assessment Steering Group. An Alaska-based official of the Minerals Management Service of the U.S. Department of Interior will serve as the U.S. co-chair. A team of international authors will address four types of impacts or effects from oil and gas activities in the Arctic:

- Social and economic consequences;
- Environmental impacts from pollution;
- Environmental effects of physical disturbances; and
- Effects on human health.

The team of authors will consider the full range of oil and gas activities. These include leasing or licensing, seismic and drilling exploration, production drilling and development construction, continuing production operations, all facets of transportation, and the eventual decommissioning of facilities. The intent of the Assessment, in keeping with the directions from Ministers, is to provide a comprehensive and balanced view of the positive and negative socioeconomic consequences associated with oil and gas development in the Arctic.

Arctic Council organizers of the Assessment plan to hold a symposium in St. Petersburg in the summer of 2005. The symposium will bring together experts in all relevant fields on the current state of the science and technology relating to oil and gas activities in the Arctic. It will be designed to promote dialogue among experts, government officials, and Arctic residents, as well as to identify critical information that may be missing from the draft assessment.

The Arctic Council's Oil and Gas Assessment illustrates how eight nations, plus observer States such as the United Kingdom, France, and Germany, and the indigenous people represented by six organizations can come together on an energy issue of critical importance. Stakeholders will have an opportunity to compare best practices in every aspect of oil and gas development – from new technologies for cleaner production, to benefit sharing with local residents, to preparing for oil spill recovery in ice conditions.

Iceland and Canada have taken the lead on another Arctic Council initiative, the Arctic Marine Strategic Plan, which will be reported out to Ministers in Reykjavik in November 2004. The United States is an active member of the drafting team. Representatives from the National Oceanic and Atmospheric Administration, the Coast Guard, and the Arctic Research Commission participate for the United States.

Ministers recognized that existing and emerging activities in the Arctic warranted a more coordinated and strategic approach to the challenges facing Arctic coastal and marine environments. Climate change and variability and increased economic activity are identified as the principal drivers of change in the Arctic Ocean and along its coasts. Scientific interest in ecosystem approaches to resource management and new national commitments to the goals outlined in the Johannesburg Plan of Implementation following the World Summit on Sustainable Development also sparked interest in the Arctic Marine Strategic Plan.

The Plan in its current form addresses four key areas: pollution, biodiversity and ecosystem integrity, communities and human health, and marine resource use. Drafters will make a strong case that regional cooperation has great potential both for addressing the opportunities and avoiding the problems of degradation at the ecosystem level.

Within the Arctic Council framework the Strategic Plan will highlight key issues for regional cooperation. The Marine Strategic Plan may also spark new interest at the national and regional levels in taking a more holistic approach to ecosystem management in the Arctic. Several studies in the Arctic identify the gaps in our capacity to understand Arctic ecosystems, particularly in coastal and marine environments. The deepening international cooperation among scientific researchers at Svalbard may offer opportunities to study one or more marine ecosystems.

The Arctic Marine Strategic Plan is not prescriptive or binding. Instead, it represents the best thinking among national oceans policymakers in the Arctic – ideas that should be shared with all who will listen.

These are but two international activities in the energy and oceans arenas. They are examples of what eight nations and other stakeholders can do together on a voluntary basis to promote sustainable development in the Arctic.

Another example of regional coordination on oceans issues is adoption of the Pacific Islands Regional Oceans Policy, another initiative developed in 2002 in conjunction with the World Summit on Sustainable Development, and the first such regional policy on oceans.[4] The Policy identifies five principles and strategic actions: improving understanding of the oceans, developing and managing the ocean's natural resources in a sustainable way, maintaining the health of the ocean, promoting the peaceful use of the ocean, and creating partnerships and promoting cooperation. It reflects the region's collective awareness of the increasing threats to the ocean environment, and the reality that sustainable economic and social development in the Pacific Islands depends on the wise use of the ocean and its shared resources.

Delegates from the Pacific Islands met in Suva in February 2004 to begin implementing the Policy by developing a Framework for Integrated Strategic Action. The Framework is scheduled to be completed by the end of summer 2004.

National governments are beginning to develop formal oceans policies, as well. Canada claims that its Oceans Act of 1997 was the first comprehensive ocean management legislation.[5] *Australia's Oceans Policy* was published in 1997 and provided a framework with goals, principles, and policy guidance, under which Regional Marine Plans are being developed.[6] New Zealand and Portugal are also working on national oceans policies,[7] as are Norway, the United Kingdom and India.[8]

In the United States, Congress directed the formation of a Commission on Ocean Policy in the Oceans Act of 2000. Its charge was to establish findings and make recommendations for a comprehensive, long-range national policy for the responsible use and stewardship of ocean and coastal resources.[9] President Bush appointed the 16 Commissioners, whose work was federally funded but was independent of any government oversight. The Commission held its first meeting in Washington in September 2001 and in the following two and a half years held nine hearings around the country and several more public meetings in Washington.

On April 20, 2004, the Commission issued its preliminary report containing more than 400 pages plus appendices. The report went to all 55 governors of U.S. states and territories, who were to submit comments on it by June 4, 2004. The Commission intends to issue its final report in September 2004. Under the Oceans Act, the Administration must submit to Congress within 90 days a statement of proposals to implement or respond to the Commission's recommendations.

The preliminary report contains a great many recommendations, but the Commission identified the following as the most critical:

- Establish a National Ocean Council, chaired by an Assistant to the President, and create a Presidential Council of Advisors on Ocean Policy;
- Strengthen NOAA and improve the federal agency structure;
- Develop a flexible and voluntary process for creating regional ocean councils;
- Double the nation's investment in ocean research;
- Implement the national Integrated Ocean Observing System.
- Increase attention to ocean education;
- Strengthen the link between coastal and watershed management;
- Create a coordinated management regime for federal waters;

- Create measurable water pollution reduction goals, and strengthen incentives, technical assistance, and other management tools to reach those goals;
- Reform fisheries management by separating assessment and allocation, improving the Regional Fishery Management Council system, and exploring the use of dedicated access privileges;
- Establish an Ocean Policy Trust Fund based on revenue from offshore oil and gas development and other new and emerging offshore uses to pay for and implement the recommendations; and
- Accede to the United Nations Convention on the Law of the Sea.

The Council on Environmental Quality is coordinating the Bush Administration's response to the Commission's report. The only appropriate comment that can be made at this time is that the response surely will heartily endorse the recommendation on U.S. accession to the Law of the Sea Convention!

The United States has accepted and acted in accordance with all the provisions of the Convention since 1983 – with the exception of Part XI. Informal negotiations to modify Part XI were launched in 1990 during the first Bush Administration and continued into 1994, when the United States both signed the Part XI Agreement and submitted it and the Convention to the Senate.

The Senate Foreign Relations Committee, under the leadership of Senator Richard Lugar, ended years of Committee inattention to the Convention by scheduling hearings in October 2003. At the first hearing, on October 14, a panel of experts on the law of the sea testified, including one of its most indefatigable and persuasive champions, John Norton Moore.

At the second hearing, on October 21, a panel of Administration witnesses from the State and Defense Departments and the United States Coast Guard testified in strong support of U.S. accession, as did a second panel of representatives of the oil and gas industry, environmental groups, fisheries organizations, and the shipping industry.

After the October hearings, Bush Administration officials worked closely with Committee staff on the draft Resolution of Advice and Consent. The Committee's effort was coordinated with great dexterity and patience by Michael Mattler. The draft Resolution contains declarations related to dispute settlement fora and subject matter, as specifically called for by the Convention. Other declarations and understandings contain interpretations of the Convention; harmonize terminology in the Convention with U.S. terminology, particularly in the area of marine pollution enforcement; and address procedural and constitutional issues.

The Senate Foreign Relations Committee favorably and unanimously reported the Convention to the full Senate on March 11, 2004.

At that point, a small but determined group of treaty opponents emerged, as Mr. Mattler explains more fully in his paper. As a result of the concerns they raised, other Congressional committees scheduled hearings on the Convention: the Senate Committee on Environment and Public Works, on March 23; the Senate Armed Services Committee, on April 8; the House International Relations Committee, on May 12; and the Senate Select Committee on Intelligence, on June 8.

Administration witnesses appeared at each of these hearings, and continued to express strong support for the Convention. They expressed the view, or rather the conviction, that accession will greatly benefit the United States by affirming and extending our sovereignty over vast resources; by protecting the navigational freedoms that are so essential to our national security; and by verifying the legal framework for protection and preservation of the marine environment that supports vital economic interests.

These witnesses acknowledged that U.S. leadership in oceans policy is gradually being undermined by our status as a non-Party to the Convention, and that the United States cannot rely on indefinite continuation of the status quo. Thus, they emphasized the need for the United States to be "at the table" of the Convention's institutions in order to influence and shape future outcomes that will affect our vital economic and security interests, including the delimitation of continental shelves. They have also responded, patiently and persuasively, to a multitude of unfounded and inaccurate assertions about the Convention's provisions.

The Senate calendar in the election year of 2004 is both short and crowded. Mike Mattler is in a better position than I to predict the chances for consideration of the Resolution by the full Senate before the end of the 2004 session of Congress; but accession is unquestionably in the best interests of the United States and of oceans policy – nationally, regionally, and globally.

Notes

[1] Burkino Faso and Morocco have recently announced their intent to accede to the Convention in the near future.

[2] Several amendments to the Convention on Safety of Life at Sea will enter into force on July 1, 2004.

[3] Information about White Water to Blue Water is available at <www.ww2bw.org>.

[4] See also the Putrajaya Declaration of Regional Cooperation for the Sustainable Development of the Seas of East Asia, December 12, 2003.

[5] Information about Canada's Oceans Act is available at <www.dfait-maeci.gc.ca/sustain/environissu/canocean>.

[6] See <www.oceans.gov.au/the_oceans_policy_overview.jsp>.

[7] See <www.oceans.govt.nz/policy> and <http://icm.noaa.gov/country/portugal/portugal.html>.

[8] Author's notes from a pre-conference working group meeting on national oceans policies at the Global Forum on Oceans, Coasts, and Islands, Paris 2003.

[9] The only comparable undertaking had been the Stratton Commission's report, *Our Nation and the Sea*, published in 1969 that resulted in the formation of the National Oceanic and Atmospheric Administration. NOAA became a center of ocean expertise, but was not the independent department in charge of all nonmilitary aspects of maritime policy that the Stratton Commission had envisioned.

The Law of the Sea Convention: A View from the U.S. Senate

Michael J. Mattler[*]

This paper will discuss views of the Law of the Sea Convention within the United States and the prospects for U.S. accession to the Convention. Though the United States played an important role in negotiating the Convention, nearly a decade after its entry into force the United States has not yet become a party to it. The Convention has been the subject of a good deal of discussion in the United States during the past year, but it remains uncertain when, and if, the United States will accede to it.

Within our system of government, treaties are negotiated by the president and then must receive approval by the Senate before the United States may accede to them. Since the conclusion of the 1994 agreement revising aspects of the Convention's deep seabed mining regime, the Convention has enjoyed consistent support from presidents of the United States. President Clinton submitted the Convention to the Senate and strongly urged its ratification.[1] Similarly, President George W. Bush's administration has expressed its support for the Convention.[2] In addition, a Commission appointed by President Bush to review U.S. policies with respect to the oceans has recommended that the United States accede to the Convention.[3]

The principal obstacle to U.S. accession to the Convention has been difficulty securing approval of the Convention by the U.S. Senate. This difficulty is not the result of widespread opposition to the Convention in the Senate. To the contrary, if the Senate had been able to vote on the Convention at any point over the past decade, there is little doubt that it would have received well over the 67 favorable votes necessary to approve it. Rather, since 1994, the Convention has been the victim of efforts by a very small number of Senators who have sought to keep the Convention from coming to a vote.

---

[*] Deputy Chief Counsel, U.S. Senate Committee on Foreign Relations. The views expressed in this paper are the personal views of the author. They do not necessarily reflect the views of the U.S. Senate Committee on Foreign Relations or its Members.

Between 1994 and 2002, the Convention languished on the docket of the Senate's Foreign Relations Committee, which has responsibility for reviewing treaties submitted to the Senate by the president and making recommendations on them to the full Senate. During this period, the Chairman of the Committee failed even to hold hearings to consider the Convention, in spite of personal pleas by Secretaries of State and Defense, senior military officials and industry groups, all of which supported U.S. accession to the Convention.

It was only in 2003 when the Senate Foreign Relations Committee came under a new chairman, Senator Richard Lugar, that the Convention finally received a Committee hearing. The Committee held two hearings on the Convention in October 2003 at which it heard testimony from representatives of the U.S. government, including the U.S. Navy; U.S. industries with interests in the oceans, including the oil and gas, commercial shipping, and fishing industries; representatives from non-governmental organizations interested in the protection of the environment; and experts in oceans law and policy, including one of the conveners of this Conference, Professor John Norton Moore.[4]

These hearings revealed widespread support for the Convention among constituencies in the United States with interests in the oceans and their uses. Witnesses emphasized that the Convention advances U.S. interests in a number of important respects:

- It contains legal protections for freedom of navigation, to which the U.S. military and the U.S. commercial shipping industry attach great importance;
- It contains rights for the United States to manage and exploit the resources of its large exclusive economic zone and continental shelf, and to define these areas with certainty, which are of importance in particular to the U.S. oil and gas industry;
- It contains binding obligations and a legal framework for the protection of the marine environment, which advances U.S. interests in ensuring the sustainability of ocean resources into the future; and
- It provides for U.S. participation in Convention institutions, including the International Seabed Authority, the International Tribunal for the Law of the Sea, and the Commission on the Limits on the Continental Shelf, which provide important opportunities

for the United States to participate in making decisions on important international oceans issues.

Following these hearings, the Foreign Relations Committee held a formal vote on the Convention in February 2004, and voted 19-0 to recommend that the full Senate approve U.S. accession to the Convention.[5]

Since the Committee's action on the Convention, a small but vocal group of critics have raised various objections to the Convention. These critics do not represent constituencies with interests in the uses of the oceans; rather, their concerns about the Convention appear to be more philosophical in nature. Most of these critics are opponents not only of the Law of the Sea Convention, but of virtually all multilateral treaties. They apparently believe that even in areas where international cooperation is necessary and where rules can be agreed that reflect U.S. interests, entering into treaties weakens the United States by constraining its freedom of action.

This general philosophical aversion to the Convention has led opponents to search for more specific reasons to object to the Convention. In this respect, a number of outlandish charges have been leveled against the Convention. Critics have charged that the Convention does the following:

- Gives the International Seabed Authority plenary authority to regulate all uses of the world's oceans and to impose taxes on U.S. companies;[6]
- Requires the United States to finance all technology and investment costs associated with deep seabed mining activities by the International Seabed Authority;[7]
- Requires the United States to seek permission from the United Nations before taking military action on the high seas;[8] and
- Envisions the establishment of a naval force under the control of the United Nations to enforce decisions of Convention bodies.[9]

The absurdity of these claims is clear to anyone who has taken the time to read the Convention.

Other charges leveled by critics of the Convention are somewhat less outlandish, but do not stand up to scrutiny. Some have expressed concern that the Convention would prohibit efforts to prevent the proliferation of weapons of mass destruction under the Bush Administration's new Proliferation Security Initiative. These concerns ignore the fact that the Proliferation Security Initiative's own rules provide that actions to be taken under the initiative will be consistent with national legal authorities and international law and frameworks, which include the Law of the Sea Convention, and that all of the United States' partners in the initiative are parties to the Convention and accordingly observe its provisions.[10]

Critics have also expressed concerns that the Convention's dispute resolution provisions are contrary to U.S. interests. Some have asserted that the Convention would allow international tribunals to pass judgment on U.S. military activities, ignoring the Convention's clear provisions permitting states to exclude such disputes from settlement under the Convention.[11] Others have raised objections to binding dispute resolution more generally, asserting that the Convention creates a risk that tribunals acting out of political motivation will issue rulings designed to constrain U.S. actions. These concerns ignore the Convention's provision for states to choose *ad hoc* forms of arbitration in most cases, rather than recourse to permanent international tribunals,[12] thereby allowing the United States to play a role in selecting the judges in such cases. These concerns also seem to assume that permanent tribunals such as the Law of the Sea Tribunal would bear some sort of institutional bias against the United States. There is no apparent basis for such an assumption beyond these critics' general distrust of international institutions.

So where does all of this leave the Convention? The next step for the Convention is for the Senate's Majority Leader, Senator Bill Frist, to schedule time for the full Senate to debate and vote on the Convention. He has declined to do so during the four months the Convention was on the Senate's calendar between the time of its unanimous approval by the Foreign Relations Committee and the time of this writing. There are a number of possible explanations for this: He may share the concerns of the Convention's critics; he may be seeking to avoid a public debate on an issue on which there is some disagreement within his party; or he may feel the Senate has more important matters to devote its time and attention to than U.S. accession to the Convention. At the time this paper was written there were only a few weeks remaining in the Senate's current session, and it remained to be seen whether the Majority Leader would schedule time for the Senate to consider the Convention prior to the end of the session. If the Senate does not act on the Convention during its current session, decisions about further action would be postponed until after the U.S. elections in November 2004, when they would fall to the president and to the leaders of the next Senate.

Further delay in U.S. accession to the Convention, of course, bears risks and costs for the United States. The Convention comes open for amendment for the first time in November 2004. If the United States is not a party to the Convention at that time, its ability to participate in consideration of any such amendments will be limited. Amendments that sought to limit navigational rights could be of great concern to the United States, particularly if they impacted the mobility interests of the U.S. military. Similarly, the work of the Commission on the Limits of the Continental Shelf is proceeding; as a non-party to the Convention, the United States is not eligible to submit a claim for the delineation of its broad outer continental shelf, which could hamper efforts to develop the resources of the shelf, including energy resources. More generally, by staying outside the Convention, the United States risks calling into doubt its commitment to the balance of interests codified in the Convention for uses of the oceans. In the long run, this could serve to undermine the order and stability on the oceans fostered by the Convention, to the detriment of U.S. interests and of all users of the oceans.

The efforts of the past year have placed the United States closer to joining the Convention than at any point in the past decade, and supporters of the Convention can take heart from the progress that has been made. More work remains to be done, however, before U.S. accession to the Convention can become a reality.

Notes

[1] *See* Message from the President of the United States Transmitting the United Nations Convention on the Law of the Sea, with annexes, done at Montego Bay, December 10, 1982, and the Agreement Relating to the Implementation of Part XI of the United Nations Convention on the Law of the Sea of 10 December 1982, with annexes, adopted at New York, July 28, 1994, and singed by the United States subject to ratification, U.S. Senate Treaty Doc. 103-39, October 7, 1994.

[2] *See* Testimony of John F. Turner, William H. Taft IV, Mark T. Esper, Admiral Michael G. Muller, and Rear Admiral John E. Crowley Jr. before the Senate Foreign Relations Committee, October 21, 2003, *reprinted in* U.S. Senate Committee on Foreign Relations, "United Nations Convention on the Law of the Sea," U.S. Senate Exec. Rpt. 108-10, March 11, 2004.

[3] *See* Preliminary Report of the U.S. Commission on Ocean Policy, April 2004, Recommendation 29-1.

[4] *See* U.S. Senate Committee on Foreign Relations, "United Nations Convention on the Law of the Sea," U.S. Senate Exec. Rpt. 108-10, *supra* note 3.

[5] *Id.*

[6] Frank Gaffney, "John Kerry's Treaty," *National Review Online*, February 26, 2004.

[7] Phyllis Schlafly, "We Should Drown Law of the Sea," *Human Events Online*, March 11, 2004.

[8] Henry Lamb, "Treaty By Stealth Again", *WorldNetDaily.com*, March 6, 2004.

[9] Gaffney, "John Kerry's Treaty," *supra* note 6.

[10] *See* Testimony of William H. Taft before the Senate Foreign Relations Committee, *supra* note 3.

[11] UN Convention on the Law of the Sea, article 298(1)(b).

[12] *Id.* Article 287.

PANEL II: INTERNATIONAL LAW AND REGIONAL ENERGY ISSUES

China's Energy Policies and the Law of the Sea

Kuen-chen Fu[*]

CONTENTS:

I. The Need for Energy Resources

In the summer of 2003, most of the provinces in China were forced to take special measures limiting electricity consumption. According to the National Electricity Network Company's statistics, the overall electricity consumption in the period from January to September 2003 reached 1,374.2 Giga kilowatt hours (Gkwh), 15.58 percent higher than what it was during the same period in 2002. The overall 2003 annual increase was about 15 percent. The increase occurred not only in coastal provinces, but also in inland areas. Consumption in every electricity network in China increased more than 10 percent.[1]

In October 2003, almost every gas station in Guangzhou and Shanghai had long lines of cars.[2] Before Christmas in 2003, even stranger things happened. The coal supply in Shanxi, the largest coal province in China, became unstable. Many local power plants were forced to cease operations.[3] These events indicate that the energy challenge in China is real and serious.

[*] Kuen-chen Fu, *a.k.a.*, Lawrence K.C. Fu, Professor of Law, Xiamen University Law School; Executive Deputy Director, Xiamen University Center for Oceans Policy and Law; Arbitrator, Taiwan, Xiamen, Wuhan, Russia (Far Eastern Region); and China Maritime Arbitration Commission (CMAC).

China exported oil for many years. However, since 1993 China has imported oil. Currently China's oil consumption increases at a rate of 6 percent annually, while crude oil production increases only 2 percent every year. Together with the country's rapidly growing economy, China's demand for imported oil has been expanding sharply. According to some observers in the late 1990s, after the year 2010 more than 50 percent of the oil consumption in China will depend on importing from the international market. It is also predicted that China's daily imported oil will exceed 7.4 million barrels, equivalent to what is consumed by all of Europe. At that time, China will become the second largest importing state of oil, second only to the United States.[4]

In reality these estimations are too conservative. In 2002, China's imported crude oil reached a new peak of 60 million to 70 million tons. That figure surged to about 80 million tons in 2003. According to the President of China Petroleum Chemical Corp., Mr. Wang Ji-ming, China might become the second largest oil importing state as early as 2004, which is much earlier than previously estimated.[5]

In 1994, the PRC government declared the Three-Gorge Project. In June 2003, a high dam of 135 meters was finished, and the first group of four generators began testing operations. As scheduled, the world's largest hydro-power plant will be fully operating in 2009. Although the project itself was severely condemned by many Chinese environmentalists, China's real need for energy has made her determined to take the risks. This is also an indicator of how badly China needs new energy resources.[6]

II. RESPONSES TO CHALLENGES

On May 30, 1997, *The People's Daily* (Overseas Edition) published an article summarizing then-Premier Li Peng's speech on China's responses to the severe energy challenges.[7] This has been one of very few elaborations on current Chinese energy policies.

In that article, Li Peng asserted that China's economy would steadily and continuously grow in the long term: "A question was raised over whether the energy industry, an important sector, can support rapid economic growth. Our answer is certain. We are confident of prospective development in China's energy industry."[8] Li Peng's confidence in 1997 was based on the following six strategies described in his speech:

1) Improving the coal industry. According to his prediction in 1997, "an average growth rate in coal production will be 2.3 percent annually from 1996 to 2000. Coal output total and export in the year 2000 will be 1.45 billion tons and 5 million tons, respectively. The share of coal in the total primary energy will decline due to technology progress, alternative energy development and restrictions in environmental protection. The elasticity of coal production growth will decrease from the current 0.38 to 0.32 by the year 2000." All of these predictions came true, but the repeated accidents in small and medium coal mines around China are making the policy makers more determined to give up those coal mines.[9]

2) Preserving oil supply stability and developing the oil industry. The strategy in the oil industry was to stabilize eastern oil fields and develop western and offshore oil fields. Premier Li Peng said, "Offshore oil and natural gas development is a result of the open-door and reform policies initiated in the [1980s]. China has cooperated with foreign companies in introducing foreign investment and advanced technology in offshore development." Li Peng emphasized that the development of natural gas in China should be moving faster. "The proved reserves of natural gas in China have been 1.3 trillion cubic meters, which provides a solid base for a possible rapid growth in the sector. It puts equal priority on developing oil and on developing natural gas. More investment will be poured into the sector and emphases will be focused on reserve exploitation, development and utilization of natural gas," he said.[10]

3) Diversification of power ownership and multiple channels of fund-raising in power development. According to Li Peng, "The strategies and policies in the electric sector are: (a) power development should fit local conditions; (b) equal emphases in both hydro and thermal power development; (c) appropriate development in nuclear power; and (d) simultaneous distribution network development.[11]

"To cope with capital availability shortages, administrations, enterprises and individuals should share costs in development and promotion of distribution systems and patch facilities. A differentiated electricity price system should be established. . . . Local governments and utilities have no right to increase electricity price rates arbitrarily. Local governments at different levels should continuously support small-scale hydropower development."[12]

4) Using nuclear power as a safe and reliable clean energy resource. Li Peng said:

> The Chinese government has selected water pressure nuclear power, which is prevalent in the world, as the dominant technology in China's nuclear power development. In 1997, China . . . installed 2.1 GW of nuclear power, accounting for only 1 percent of the national total capacity in electricity generation. A new total of 6.6 GW capacity in nuclear power will be added by the year 2005, accounting for about 2 percent of the national total by that time. In the short and medium term, nuclear power plays a very small part in the power sector. However, nuclear power is prospective in the long term.[13]

5) Equal emphasis on both energy development and conservation and higher priority on energy conservation in the short term. This is a critical part of China's energy policy. With economic growth and a promotion in living standards, energy consumption has certainly increased. One way to keep economic growth is to consume less energy and produce more.

As Li Peng said in his speech, in the 10 years between 1987 and 1997, China had made remarkable achievements in energy conservation. The elasticity of energy consumption declined from 0.59 to 0.39, equivalent to 470 million tce (ton coal equivalent) of accumulated energy savings. "However, compared with advanced levels of energy efficiency in the world, China is still far behind. Currently [in 1997], the overall energy efficiency in China is about 32 [percent], 10 percent lower than the developed countries in the world. GDP per unit energy consumption in China is about one half to one third of developed countries. Obviously, there is great potential in energy conservation in China."[14]

6) Energy development and sustainable development strategy. Li Peng said:

> A great deal of energy consumption pollutes the environment and damages ecological systems. The coal used in boilers and stoves is a major source of CO2 emissions and dust particles. Exhausted gases from vehicles also pollute the air. Combustion of both high sulfur content coal and oil results in acid rain. Coal ash and waste solids in boiler combustion damage water resources and lands.

> Coal and oil development brings some negative impacts on the environment surrounding the coal mines and oil fields. Therefore, we should emphasize treatments for the pollutants accompanying energy development and utilization. The policy that concurrently links energy development, energy uses and environmental protection/treatment must be implemented.[15]

He also pointed out the following:

> In energy development and utilization we should promote what is beneficial and abolish what is harmful. We should protect the environment and ecological systems and implement sustainable development strategies. We should not only consider energy uses for this generation, but also reserve sustainable energy resources for the next generation. We believe that new energy resources will be discovered and developed with progress in technology and science.[16]

With the above mentioned energy policies, Li Peng's confidence in 1997 was proven to be realistic as the country's Gross Domestic Product (GDP) grew by 9.6 percent between 1979 and 1999.[17] However, the demand for energy in China has been growing at the same speed. In addition, more sophisticated power supply management is an issue. One of the key reasons for the 2003 electricity shortage in China was the lack of a reliable estimation of supply and demand and the absence of an effective emergency reaction plan.[18]

The basic difficulty now faced in China, as Li Peng correctly described in 1997, is still the cruel fact that China has limited energy resources. Though it is the largest coal supplier, Datong Coal Group in Shanxi Province is trying to establish a market-sensitive new coalition; the Zhejiang and Hunan provinces are reviving some local thermal power plant projects; and the Hubei Province is pushing seriously the third Chinese nuclear power plant project.[19] Members of the Chinese central government are pondering the basic questions of how to maintain sustainable development and where to find sustainable, renewable energy resources for future generations.

III. Renewable Energy Resources in China

In terms of meeting energy needs in the 21st century, many developed countries have made renewable energy a cornerstone of their energy policies. The European Union has publicized its white paper on energy issues, and set the goal of having 12 percent of the E.U.'s electricity generated by renewable energy by 2010, and 50 percent by 2050. China has also signed a number of related international treaties, including the Rio Declaration and the Framework Convention on Climate Change. The Chinese government published China's "Agenda 21" and developed the "Ten Major Responsive Strategies on China's Environment and Development." Though some steady growth has occurred, the actual implementation of these policies on renewable energy has lagged far behind.[20]

In coping with the severe situation, the People's Congress of China adopted a legislative plan to have a new "Law on the Utilization of Renewable Energy" enacted and promulgated in 2005. According to a report, Mr. Li Jun-feng, Deputy Director of the Energy Institute of the China National Development Reformation Commission, said that a 2003 energy shortage made the People's Congress determined to reschedule its five-year legislative plan and bring that bill from the back burner to the front. Li pointed out that the new law would authorize the government to take responsibility for promulgating strategic plans and safeguard measures, ask the citizens to pay the extra cost for the renewable energy development, and provide for a scheme for encouragement and sanctions.[21]

Renewable energy development is particularly important for China for the following reasons: (1) energy security; (2) international competitive pressure; (3) health and environmental impacts of the fossil-fueled economy; and (4) economic impacts.[22] According to some experts' analysis, in some technologies, notably solar water heating, China has a well-established industry with decent-sized export markets. Other technologies are also rapidly moving towards commercialization. These include the following:[23]

A. Commercialized:

(1) small hydropower;

(2) solar water heating;
(3) passive solar house;
(4) solar stove;
(5) geothermal powcr wind turbines;
(6) geothermal heating;
(7) traditional bio-energy technology; and
(8) small methane tank.

B. Early commercialization:

(1) solar cell;
(2) grid-connected wind turbine;
(3) small and mini wind turbines;
(4) large and medium methane tech; and
(5) tidal power generation.

There are other renewable energy technologies, which are still in the demonstration stage or even the R&D stage. These include the following:[24]

C. Demonstration:

(1) solar dryer;
(2) municipal organic waste power generation; and
(3) biomass gasification.

D. Research and development:

(1) modern bio-energy techniques other than the above mentioned bio-energy techniques;
(2) wave power generation;
(3) ocean thermal energy conversion;
(4) new hydrogen manufacturing technologies; and
(5) hydrogen storage techniques.

IV. Asking for Energy From the Sea

For the purpose of proliferation of energy resources, and for the purpose of finding sustainable energy resources, the Chinese are going more aggressively to the seas. With a coastline 18,000 km long, China has begun using wind power and solar power along the shore in the coastal areas.

Other than the wind energy and solar energy technologies listed previously, the other main energy resources from the sea should include: (1) wave power; (2) tidal power; and (3) ocean thermal energy. Though tidal power has been utilized in China for almost 1,000 years since the Tang Dynasty for pushing some small coastal mills, it has never been used widely in this part of the world.[25] Since technologies for these three ocean energy resources are still in the research and development stage in China, exploitation of these energy resources at sea has not been commercialized; and therefore they are not yet reliable sources for China's energy today. China still has to look eagerly at the country's only reliable source of energy from the sea – the oil and gas resources in continental shelves.

China began developing a modern oil and gas industry in the 1950s. In 1952, the Ministry of Geology and Mineral Resources was established. Based on Soviet experts' suggestions, a survey team explored for oil in Gansu, Xinjiang, and Qinghai in northwestern China. After a partial investigation, Soviet experts concluded, as did western geologists, that China was an oil-poor country.[26]

In 1954, Li Siguang, Director of the Ministry of Geology and Mineral Resources, led a new team organized to conduct a strategic oil investigation and exploration throughout the nation. The successful development of the Daqing Oil Field in the 1960s made China self-sufficient in oil. Since the 1970s, China has explored and developed the Shengli, Dagang, Liaohe, Jianghan, Huabei, Changqing, Jilin, Zhongyuan, Henan, Jiangsu, and Jidong oil fields. The Sichuan gas field has also been further developed.[27]

China's oil and gas production had increased steadily each year, though it has slowed down in recent years. In 1996, China's crude oil production set a new record of more than 156 million tons, which is eight million tons more than in 1995. The gas output in 1996 was 16.44 billion cubic meters, 290 million cubic meters more than in 1995. At the same time, offshore oil production also increased rapidly and accounted for 81 percent of China's yearly increase in crude oil production.[28]

China National Offshore Oil Corporation (CNOOC), established in 1982, is responsible for developing China's rich offshore oil and gas reserves. CNOOC explores and develops offshore oil and gas and processes and sells these products alone and jointly with foreign entities. CNOOC is a well-equipped, highly efficient international petroleum corporation with a staff of 30,000. The State Council closely supervises the development of China's offshore petroleum industry. It has also allowed CNOOC to build power stations, fertilizer plants, refineries, and other industries that are heavily dependent upon petroleum resources, and to purchase offshore oil and gas at international prices for use in these industries.[29]

CNOOC has built a fertilizer plant that produces 2.5 million tons of urea by using gas. It has plans to build two petroleum plants in Huizhou, Guangdong Province to refine 8 million tons of oil and produce 450,000 tons of ethylene. By the end of 2010, China's crude oil processing capacity will have reached 300-350 million tons. Its ethylene production capacity will be at 8-10 million tons, and the petrochemical industry will have developed to an advanced international level.[30]

Given the shrinking oil reserves on land and slow increases in output from old land oil fields, China's offshore petroleum industry is being rapidly developed. It can be said that China's oil increases are largely provided by increased offshore crude oil. The offshore petroleum industry is becoming increasingly important to China's national economy. In the 1960s and 1970s, China had been confined to traditional production and exploration methods. By the end of the 1970s, China had developed only three offshore oil fields covering three million square kilometers.[31] China's offshore petroleum industry began developing rapidly after CNOOC was established in 1982. From 1982 to 1996, its annual output of offshore crude oil increased from 80,000 tons to 18 million tons. CNOOC has helped stabilize China's domestic oil and gas supplies. In 1996, its offshore output of oil reached 15 million tons, and gas reached three billion cubic meters. China's 13 offshore oil and gas fields have profit margins of 40 percent. CNOOC has developed 19 large offshore oil and gas fields to modern, international standards. It can successfully develop huge, offshore, thickened oil fields with reserves in excess of 100 million tons as well as small, offshore oil fields with reserves of only several million tons.[32]

V. Development of Offshore Oil and Gas Fields in the China Seas

China has 1.3 billion tons of proven offshore crude oil reserves and 200 billion cubic meters of proven offshore gas reserves. The Bohai Bay, which is sometimes called the Bohai Sea, alone has proven oil reserves of 600 million tons, or half of the nation's total offshore reserves. One medium-sized oil and gas field discovered in the inter-tidal zone of the Bay of Bohai Sea has geological reserves estimated at 50 million tons.[33]

The first phase of China's largest offshore oil field, the "Penglai 19-3" oil field, has commenced operations, according to officials with Bohai oil field. Located in the southern part of the Bohai Bay, the oil field has a tectonic acreage of 50 square kilometers and a proven reserve of about 600 million tons, the largest ever discovered offshore in China. The oil field was first detected in 1999, through joint efforts by the Phillips Petroleum Company of the United States and the CNOOC. After careful research and studies, construction of the first phase, which consisted of one platform and 20 oil wells, began in 2001.[34]

According to Zhou Shouwei, President of CNOOC, the start-up of operations of the first phase, with a designed daily production capacity of 40,000 barrels, marks the start of Sino-U.S. oil exploitation cooperation in the Bohai Bay. Upon completion of the project, nearly 200 oil wells are expected to produce an annual average of more than 10 million tons of crude oil.[35]

Development of oil and gas fields in Bohai Bay has also brought new breakthroughs in the field of ocean management in China.[36] "The PRC Coastal Water Areas Utilization Management Law" was promulgated on October 27, 2001, and entered into force on January 1, 2002. According to Article 3 of the law, coastal water areas are solely owned by the State. The State Council is the representative of the State for exercising the latter's ownership. Any utilization by any individuals or entities shall be subject to permits made by the State Council or its authorized government agencies.

After serious efforts made by the State Ocean Administration (SOA) of the State Council, CNOOC eventually submitted applications in May 2003, asking for permits to continuously use the oil fields in the Bohai Bay. However, the applications for use of the eight CNOOC oil fields, together with the pertaining feasibility reports, were rejected by the SOA. More detailed information and specific application forms for the platforms and submarine cables were requested by the SOA. In accordance with the "Technical Rules for Coastal Water Areas Management," the SOA decided the outer limits of the CNOOC's used water areas in the Bohai Bay, based on the area used by each platform and each submarine cable. CNOOC has re-submitted its applications to the SOA with improved forms for all eight Bohai Bay oil fields, 32 platforms (including storage tanks) and 48 cables, and with improved feasibility reports. It is expected that after reviewing these documents, permits will be given to CNNOC for its continuous usage of these water areas in the Bohai Bay.[37]

The South China Sea area is also a promising area for China's offshore oil and gas development. According to Chinese experts' explorations, gas reserves in the South China Sea area are estimated (not proven) to be 13 trillion cubic meters; oil reserves are estimated to be 10.2 billion tons. Oil and gas reserves are concentrated in the offshore waters around Hainan Island. Four oil and gas bearing structures have been discovered there: the Beibuwan, Yinggehai, Qiongdongnan, and Zhujiangkou basins.[38]

Good news has come from oil and gas explorations in these offshore waters around Hainan Island. The South China Sea Petroleum West Corporation (a subordinate of CNOOC) drilled a test well, "Oriental 1-1-5," and gained productive oil and gas flows in the Yinggehai Basin to the west of Hainan Island. This supports the opinion of experts that the "Oriental 1-1" structure alone contains geological gas reserves of 40-60 billion cubic meters.[39]

The Liuhua 11-1 oil field is, to date, the largest oil field in the South China Sea and the reservoir is located in 1,000 ft (300 m) of water. After preliminary exploration studies, Liuhua 11-1A was drilled. It tested at rates of 2,240 barrels (bbl) of heavy crude oil per day. Confirmation drilling of two additional wells, the Liuhua 11-1-3 and the Liuhua 11-1-4, showed a significant reservoir accumulation of more than one billion barrels of oil-in-place.[40]

Because of the oil's very heavy, dead crude characteristics and the relatively shallow depth of the reservoir, it was clear from the initial discovery that innovative technology would have to be combined with proven techniques to bring Liuhua to life.[41]

Between 1987 and 1992, Amoco and Nanhai East (another CNOOC subordinate) engineering teams experimented and tested varying development concepts until they found a breakthrough solution. The teams determined that the offshore production system would have three major components:

(1) A floating production system (FPS) for drilling and producing well support;
(2) A floating production, storage and off-loading (FPSO) system for processing and storing the crude oil; and
(3) An innovative subsea system using electric submersible

pumps to service approximately 20 horizontal wells.[42]

In March 1993, six years after the discovery, the overall development program for the Liuhua 11-1 field was approved by all levels of the Chinese government. The project was eventually put on a fast track leading to its success.[43]

In the East China Sea area, as reported on August 20, 2003, leading domestic and foreign oil explorers have clinched multi-billion-dollar deals to accelerate the development of oil and gas projects. Five prospecting and exploitation contracts were signed between CNOOC, the China National Petrochemical Corp (CNPC), the Shell Group and Union Oil of California.[44]

The two Chinese companies hold 30 percent of the rights and profits under the agreement, and the foreign companies take the remainder. CNOOC will run the operations, which are situated east of China's economic hub, Shanghai, and the province of Zhejiang. Prior to the contracts being inked, Premier Wen Jiabao met the senior executives of the four groups, stressing that close cooperation between Chinese and foreign companies will accelerate the development of oil and gas exploration in the East China Sea. Wen also said he hopes such cooperation can help guarantee the energy supply for the booming Yangtze River Delta. To cover the energy demands in the economically developed East China region, the country has already launched projects in recent years to transport electricity, oil and gas from the nation's western regions.[45]

An official from the National Development and Reform Commission said that the contracts cover exploration, development, production, transportation and the sale of oil and gas from the region: "This is the biggest oil and gas-related project signed between China and foreign countries during recent years." The official said the cooperation will speed up exploration in the area, as previous efforts had only been carried out on a relatively small scale. Experts estimate that about 28-38 billion cubic meters of gas and 10-20 million barrels of oil can be extracted from the area under the agreements.[46]

VI. External Cooperation on Offshore Oil and Gas Exploration

Extracting offshore oil and gas is an expensive, high-risk operation that involves advanced science and technology. The offshore petroleum industry was the first industry in China to open to the outside world. External cooperation began in 1982 when CNOOC was founded.[47]

Through four rounds of public bidding and negotiations, CNOOC signed 119 oil contracts with 65 foreign companies from 18 countries. It has absorbed foreign investments of US $5 billion, accounting for 58 percent of offshore exploration investments. Altogether, more than 400 offshore oil wells have been drilled and more than 90 oil- and gas-bearing structures have been discovered.[48]

CNOOC has undergone three stages: It went from being simply a source of raw materials to being a partner with large western petroleum corporations, to being an unprotected international competitor dependent solely upon its own strength and international reputation.

In 1994, 30 days ahead of schedule, the Bohai Sea Offshore Drilling Rig No. 4 drilled a test well to a depth of 3,800 meters in Japan's Bay of Ishikari. This was a historic moment in overseas drilling for China's offshore petroleum drilling industry. In 1995, Bohai No.12 started drilling more oil wells in Japan's Akita waters. China's petroleum industry has established a model of external cooperation with Chinese characteristics.[49]

Since the 1980s, China's offshore petroleum industry has made great achievements through cooperation with transnational corporations such as the French company TOTAL and the American corporation Atlantic Richfield Company (ARCO).[50] China's oil and gas industry has a tradition of self-reliance. After adopting the policy of reform and opening, it has absorbed foreign investment, and adopted advanced foreign technology and management expertise. With increased Sino-foreign exchange, China's oil and gas industry has begun joint exploration and development of small- and medium-sized overseas oil fields. China is also working to join large cooperative projects.[51]

On January 1, 1994, CNOOC bought all of the shares owned by the American company ARCO in the Malacca contractual area of Indonesia. This was the first overseas contractual area for CNOOC. It was also the largest and the most profitable overseas project in China's petroleum industry. China's petroleum department has been exploring and developing oil and gas fields in Peru, Thailand, Canada, Sudan, Papua New Guinea and elsewhere. It has also become involved in bidding projects, one being the development of a large oil field in Kazakhstan.[52]

The cooperative work between China and the Gulf States is also ongoing. Saudi Arabia plans to build an oil refinery joint venture in Shandong Province. CNPC has signed a contract with Kuwait to produce U.S. $400 million worth of oil equipment. In addition, Chinese oil companies are seeking joint cooperation in drilling other oil fields in the United Arab Emirates and other Gulf States.[53]

VII. New Regulations for External Cooperation on Offshore Petroleum Exploitation

The success of China's exploration and exploitation of offshore oil and gas resources since 1982 has been partially based on rather successful regulative works on the offshore petroleum projects. The general comment on the 1982 Regulations of the PRC on the Exploitation of Offshore Petroleum Resources in Cooperation with Foreign Enterprises (hereinafter referred to as "the 1982 Regulations") is that it has offered a set of rational, modern rules, reflecting the most flexible norms for foreign enterprises doing business with CNOOC.[54]

According to the 1982 Regulations, a complex form of petroleum cooperation agreement was adopted in China for CNOOC and its contracting partners. Foreign contracting enterprises were taking risks for their own investment. They were the real operators of the offshore oil fields in every stage of the exploration and exploitation procedures. Foreign contracting enterprises had to join in a non-juristic person joint management committee together with CNOOC for offshore exploration and exploitation activities, though CNOOC had the final say on several substantial matters. Foreign enterprises obtained revenue from the petroleum extracted from their concession areas. They obtained both ownership and the right of dispensing petroleum in accordance with the ratio provided for in the CNOOC contracts. All data, samples, records and/or other original materials obtained by the foreign contracting enterprises during their operations in the concession areas should eventually be transferred to CNOOC.[55]

On September 23, 2001, the 1982 Regulations were amended and re-promulgated. The 2001 Regulations continued the tradition of a single-windowed cooperation model. The CNOOC is still the only window for external cooperation on the offshore oil and gas development. However, some main differences may be observed from the new 2001 Regulations:

1) The department designated by the State Council has taken the place of the Ministry of Petroleum Industry as the competent authority to determine the forms of cooperation, to demarcate areas of cooperation in accordance with the zones and the surface areas of cooperation, and to approve the master projects for offshore oil and gas field development.
2) The Ministry of External Trade and Economic Cooperation, instead of the Foreign Investment Commission, is designated by the new Regulations to approve contracts signed between CNOOC and foreign enterprises. (Since 2003, the newly established Ministry of Commerce has substituted the Ministry of External Trade and Economic Cooperation, because the latter was merged into the new Ministry of Commerce.)
3) In the 1982 Regulations, standardized terms and conditions of the petroleum cooperation agreement as provided in Article 7 were allowed to be modified only by the Ministry of Petroleum Industry's order or the [CNOOC] contract. The new 2001 regulations allow such modifications by "other laws, regulations, or the [CNOOC] contracts."
4) Instead of being in accordance with the provisions of the Interim Regulations on Foreign Exchange Control of the People's Republic of China, foreign contractors shall be "in accordance with the PRC's Regulations on Foreign Exchange Management and any other pertaining rules" when opening a bank account or handling foreign exchange matter.
5) The 1982 Regulations provide that "in petroleum operations, the foreign contractor must give preference in employment to Chinese personnel, progressively increase the percentage of Chinese personnel and train Chinese personnel in a planned way." The 2001 Regulations have lessened the obligation and provide that foreign contractors "may" employ with preference the Chinese

citizens for their operations in China.

6) The 1982 Regulations state that CNOOC shall have the right to send personnel to join the foreign operator in making the master designs and engineering designs for carrying out the petroleum contract. Designing corporations within the territory of the People's Republic of China shall enjoy priority in entering into subcontracts for the master designs and engineering designs mentioned above, provided that their terms are competitive. The new Regulations, however, have deleted the second part of the original provision. Because CNOOC still has the right to participate in the design works done by the foreign contracting parties, the designing companies within the PRC territory no longer have the privilege of entering into subcontracts for the master designs and engineering designs.
7) Also deleted from the 1982 Regulations are the provisions of Articles 19, 20 and 21 concerning privileges enjoyed by competitive companies within the PRC territory to offer facilities, equipments, constructions, materials, and services needed for the operations.
8) The 2001 Regulations also provided a modernized provision concerning requisition and compensations. The original 1982 Regulations provided in Article 26 that in circumstances of war, the threat of war or other emergencies, the Chinese Government shall have the right to compulsory purchase or requisition of some or all of the petroleum due to a foreign contractor or purchased by it. The PRC government promised that China will not requisition the investment or revenue of a foreign enterprise, which takes part in the offshore petroleum development cooperation. Under special circumstances, based on needs of social public wealth, the Chinese government may requisition a portion or all of the petroleum due to the foreign enterprise participating in the offshore petroleum resources development. In that case, appropriate compensation shall be given to such foreign enterprises. Such a change can also be noticed in the 2004 Constitution amendments. Requisition without

appropriate compensation is not unconstitutional in the PRC.

9) "The department designated by the State Council" is also authorized in the new 2001 Regulations to warn and to stop, if necessary, the operators or contractors in violation of the Regulations. Violators will be liable for all economic damages.

VIII. Issues Ahead and the Law of the Sea

China not only ratified the United Nations Convention on the Law of the Sea (UNCLOS) on May 15, 1996, but also promulgated the Law on the Exclusive Economic Zones and Continental Shelves on June 26, 1998. The country has also enacted and promulgated many other detailed statutes on territorial sea, marine environmental protection, fisheries management, maritime zones utilization, underwater cultural relics preservation, and offshore oil and gas resources, etc. In terms of establishing a regime of pertaining laws and regulations, China is in a leading position. Nevertheless, this does not mean that China is not having difficulties with exploration and exploitation of the hydro-carbon resources from the sea. The many issues remaining for China's endeavors to retrieve energy from the sea are mainly political issues subject to equitable solutions based on the Law of the Sea. Boundary delimitations and joint development projects are among the most urgent issues.

China has not yet submitted its outer continental shelf limits to the United Nations Commission on the Limits of Continental Shelf (CLCS). Once China submits these to the CLCS, it is expected that some disputes will arise with neighboring states and other members of the international community. For decades, China has postponed maritime boundary issues for the number one national policy, i.e., peaceful reformation and opening up. Eventually, however, maritime boundary delimitation issues will have to be determined by China and its neighboring states.

Since China needs the oil and gas resources from the sea, it seems reasonable to adopt the policy of "maintaining sovereignty, postponing disputes, and developing jointly." China has used this as a means of maintaining peace with neighbors in recently years. However, developing jointly has not been successful. One of the reasons, in this author's opinion, is that China's neighbors are obviously more interested in developing without China's participation than in delimitating any joint development zones at sea. Particularly in the South China Sea area, states occupying the Chinese-claimed islets in the Spratly Islands are trying to extract oil and gas before any legal arrangements are made with China. They are not interested in making boundaries or making any joint efforts to share resources with China in the disputed areas.

Part XV of the United Nations Convention on the Law of the Sea (UNCLOS) has provided some procedures for settling this type of dispute. Article 287 has provided the following procedures for the contracting party states to choose:

> (a) the International Tribunal for the Law of the Sea (ITLOS) established in accordance with Annex VI;
> (b) the International Court of Justice (ICJ);
> (c) an arbitral tribunal constituted in accordance with Annex VII; and
> (d) a special arbitral tribunal constituted in accordance with Annex VIII for one or more of the categories of disputes specified therein.

It remains to be seen whether China will be forced to use any of these procedures to settle the disputes with its neighbors. What is certain is that if China decides to use these procedures, the country will be prepared with sufficient arguments based on not only the geographical and geological reasons, but also the vested rights and historic titles.

Notes

[1] Cui Yi, "Historical Challenges for the Energy Chain in 2003," *Zhong-guo-jing-ying-bao* (*China Management Daily*), <http://www.gx.cei.gov.cn/list.asp?id=29532> (as shown on 14 May 2004).

[2] *Ibid.*

[3] *Ibid.*

[4] Wang Qiao-ping, Anti-terrorism and China's Security Policies, *Zhong-guo-ping-lun (China Review)*, Vol. 76, April 2004, p. 92.

[5] Dong Wen, "2004: Energy Watch points," <http://www.xzdc.cn/ArticleShow.asp?ArticleID=270> (as shown on 14 May 2004).

[6] Electricity generating began from July 10, 2003. By October 22, 2003, the Three Gorge Project generated 3.572 Gkwh. Electricity generating in year 2004 is estimated at 30.9-35.2 Gkwh. *See, ibid.*

[7] <http://www.pnl.gov/china/lipeng.htm> (as shown on May 14, 2004).

[8] *Ibid.*

[9] *Ibid.*

[10] *Ibid.*

[11] *Ibid.*

[12] *Ibid.*

[13] *Ibid.*

[14] *Ibid.*

[15] *Ibid.*

[16] *Ibid.*

[17] *China Statistics Yearbook 1999*, State Statistics Bureau, China Statistics Publishing House, September, 1999.

[18] Another reason for the 2003 electricity shortage was probably grounded on the theory that during the 1997-1998 Asian financial crisis, the Chinese authorities stopped all the

new thermal power plants projects for 1999-2002. This, if it is real, is obviously another indication for the need of better prediction capabilities and more sophisticated management. *See*, Cui Yi, "Historical Challenges for the Energy Chain in 2003," *Zhong-guo-jing-ying-bao* (*China Management Daily*), <http://www.gx.cei.gov.cn/list.asp?id=29532> (as shown on 14 May 2004).

[19] *Ibid.*

[20] Zhang Zheng-min, Wang Zhuang-yi, et al., *Renewable Energy Development in China: The Potential and Challenges*, Coal Industries Publishers, November 2002, p.81.

[21] "Legislative Plan Adjusted: Renewable Energy Utilization Law Expected Be Promulgated Next Year", *Beijing Morning Paper*, 25 May 2004, p.1.

[22] Zhang Zheng-min, Wang Zhuang-yi, et al., *Renewable Energy Development in China: The Potential and Challenges*, Coal Industries Publishers, November 2002, p.81.

[23] *Ibid.*, pp. 84-85.

[24] *Ibid.*

[25] <http://www.fsyz.com.cn/xuexiao/feshan/dili_sea/hanyannenyan/hyny55.htm> (as shown on May 14, 2004).

[26] Source: China's Petroleum Industry, China News Organization, <http://www.vitrade.com/china/chinanews_brieing_oil_industry.htm> (as shown on May 14, 2004).

[27] *Ibid.*

[28] *Ibid.*

[29] *Ibid.*

[30] *Ibid.*

[31] *Ibid.*

[32] *Ibid.*

[33] *Ibid.*

[34] *People's Daily*, Last updated at: (Beijing Time) Thursday, January 09, 2003, <http://english.peopledaily.com.cn/200301/09/eng20030109_109849.shtml> (as shown on May 14, 2004).

[35] *Ibid.*

[36] <http://www.soa.gov.cn/>as shown on 2004-5-10.

[37] *Ibid.*

[38] Source: China's Petroleum Industry, China News Organization, <http://www.vitrade.com/china/chinanews_brieing_oil_industry.htm> (as shown on May 14, 2004).

[39] *Ibid.*

[40] <http://www.offshore-technology.com/projects/amoco/> (as shown on May 14, 2004).

[41] *Ibid.*

[42] *Ibid.*

[43] *Ibid.*

[44] *People's Daily*, Last updated at: (Beijing Time) Wednesday, August 20, 2003, <http://english.peopledaily.com.cn/200301/09/eng20030109_109849.shtmldaily.com.cn/200308/19/eng20030819_122647.shtml> (as shown on May 14, 2004)

[45] *Ibid.*

[46] Alexander's Gas and Oil Connections, *Company News E and SE Asia*, volume 8, issue #17 - Thursday, September 04, 2003, adopted from original source: *China Daily*, <http://www.gasandoil.com/goc/company/cns33616.htm> (as shown on May 14, 2004).

[47] Source: China's Petroleum Industry, China News Organization, <http://www.vitrade.com/china/chinanews_brieing_oil_industry.htm> (as shown on May 14, 2004).

[48] *Ibid.*

[49] *Ibid.*

[50] *Ibid.*

[51] *Ibid.*

[52] *Ibid.*

[53] *Ibid.*

[54] Yu Min-cai, *Hai-yang-shi-you-kan-tan-yu-kai-fa-de-fa-lv-wen-ti* (*Legal Issues for the Exploration and Exploitation of Ocean Petroleum*), Chinese People's University Press, 2001, pp. 45-57.

[55] *Ibid.*, p.56.

Problems of International Law Legislation for Exploration and Transportation of Energy Resources by the Sea in Light of Terrorist Activity

Boris A. Smuslov[*]

"Navigare necesse ist, vivernon est" – One must live to navigate, life is a second quality. These words attributed to Pompeius in 67 B.C. were sung in the teeth of Roman ship captains who, fearing the hand of pirates, ventured to follow the senate's order to put out to sea.

The utterance could easily be paraphrased today as "One must navigate to live." Seafaring has always been a risky business. Absolute security was in fact unachievable, as captains were unable to control conditions. It is not only storms, hurricanes and wrecks that destroy seamen's lives and health. More often it is terrorists and modern pirates who attack seafaring people.

Mankind has always been deeply interested in the sea. It is of a global and lasting character as a result of various countries' requirements: sovereign power, territorial reconnaissance, capture of living and inanimate resources, location of some particular infrastructure's elements, protection of sea surroundings and the environment, realization of scientific research, utilization of the Pacific for the international cooperation in economics and other spheres evoking mutual interest. But the ocean is not only utilized for the benefit of mankind. Actions are being committed that pose a threat to every state's interests.

The most dangerous threats are considered criminal actions by international law and customary standards. A few of these threats are committed by individuals and are not associated with the activities of a country. Social relations of the countries included in the sphere of state preservation of law and international relations are regarded as objectives for encroachment. One uses the term "criminal actions of international character" to define these crimes in the theory of international law and in the practical activity of international association.

Where crimes are committed on ships or against ships or against the state's objectives on the sea, the crimes are defined by international law. Criminal actions at sea include actions threatening sea navigation safety, static platforms situated on the continental shelf and piracy. It should be

[*] Maritime Law Chair, Maritime State Technical University.

noted that some of those international crimes committed on the sea are applicable just for the sea surroundings.

A special feature of sea terrorism is that sometimes one cannot distinguish between terror preparations and the terror act itself. An example is collecting money and preparation of technical means, which can be done by piracy.

The objective of terror attacks on the sea can be to exterminate important navigable ways or to curtail sea tourist business. The most dangerous is damage to international transportation routes, especially routes for raw material. Attacks on oil platforms influence social opinions by telling the public about an impending ecological catastrophe. This can be done with the aid of aiming at a port operation using accessible mine weapons or easy divisional means. Because this is achievable, it is very difficult to control.

Terrorism has always existed and apparently it will continue to exist out of righteous or unrighteous anger, frenzy, despair or a sense of violated justice, fanaticism, hatred, deadlock or fits of madness. Yet the weapons are not the same. At one time knives, revolvers and hand grenades were used, while today terrorists are using the latest technological devices.

The specific features of international criminal action at sea are now being reflected in international contracts. The signs are emphasizing the special character of the sea and its general international features.

The given signs reflect the peculiarity of international legal status and the legal conditions of definite sea territories within which criminal actions occur and the specific features of the crimes. The increased social danger is the most impotent social (material) sign of the crimes. As for the priority of the defense from criminal danger of an international character, the world gives prominence to a man, his life, health and a lot of generally accepted rights and liberties.

The sea specific character of the crimes is often inserted based on instructions about the kinds of transportation used by criminals in international contracts and also of their specific features that are sometimes signs of violations of international law.

Persons who commit these criminal actions bear the responsibility for them and are to be punished in conformity with international law standards or state laws. International law defended interests, also interconnection of the signs of the citizenship of the people who commit these criminal affairs and also the sea territories where the latter took place, point out the concrete art of crimes indicated in international law.

International sea terrorism has its own specific features. One of them is the subject of the crime. The equipment and armament of sea terrorists is more complicated and expensive than those of the persons who commit crimes on land and in the air. Such activity presupposes that sea terrorists cannot act individually but act in groups supported by a government, sovereign people or even organizations that are interested in their activity.

Sea terrorism menaces the international security of high seas navigation and violates the general principle of freedom of the seas.

The first wave of international terror that was referred to by Western critics in 1968 flashed in many countries in numerous criminal actions, mainly on the airlines and in air crafts. After governmental and commercial air cooperation had taken all due measures of safety in defending civil airplanes, the terrorists began to attack on land: airports, legations, and some individual great business agents and political establishments that did not remain without giving attention to the part of most of the countries' state security.

The contraction in counteracting terrorism on land and in the air, especially after the attacks in September 2001 on the buildings of the World Trade Center, force the terrorists to search for new, less defended objectives to attack.

As many foreign commentators see it, at the present time and in the very near future, international terrorism will be concentrated on the more accessible sea aims such as commercial lines of communication, ocean transport and passenger ships.

The vulnerability of naval and commercial shipping to sea terrorism attacks was quite convincingly demonstrated on October 12, 2000, when members of terror group al Qaeda organized an attack on the American destroyer Cole, which stood in the Yemen naval port Aden.

After the end of the Cold War, sea terrorism and pirate actions have been compromising new world regions. The terror attacks have become more intensive and severe, cause considerable material loss and casualties. The attacks also cause serious damage to sea trade and transport shipping, as well as to Navy ships and the Coast Guard of a great number of countries. According to Western data, about 90 percent of sea terror crimes took place in the water territories adjacent to the coasts of developing countries, and governmental structures have proven to be ineffectual, poorly financed, often corrupt and incapable of fighting against terrorism, which is becoming better organized. Thus, the organization fighting against sea terrorism and all of its manifestations draws the attention of the civilized world.

The important part in this field that law plays is in measures influencing directly the effectiveness of the fighting on the whole. An analysis of multilateral international contracts in this field affirms that these agreements have only general bases of international cooperation. In the contractual practice of states in this field one must move on to the concrete measures providing for the sea-specific features of this fighting.

There are plenty of international law acts and a number of organizations with different standards that take an active role in the war against international terrorism or help to coordinate against this fight. There is, however, no universal international contract at present that defines the terms "international terrorism" and "international sea terrorism" and their *corpus delicti*. There is no easy interpretation of the term, and the opinions of leading experts on this matter are not the same. In other words, a point of view on the basis of which a unified legal machinery could be built does not exist.

When talking about universal international contracts connected with the sea-specific feature, one must not forget the Convention for the Suppression of Unlawful Acts against the Safety of Maritime Navigation and the Protocol for the Suppression of Unlawful Acts against the Safety of Fixed Platforms Located on the Continental Shelf (Rome, 1988). And at the same time, it could be a false opinion that the Convention of 1988 and the Protocol of 1988 are dedicated to the fight against international terrorism. It is a question of whether both terror and other purposes have the same dangerous consequences. These treaties directly concern the international law of ship security and port means. This paper will not analyze these documents, as it has been done before.

Preventing terrorist acts and stopping legal actions against shipping must be coordinated with international law, in particular the international law standards in the field of Human Rights. One should also try to employ more strict means of security while supporting international trade.

New measures of safety are already telling on the industry of sea transportation. For example sea insured duties have raised three fold for the tankers crossing Yemen waters. The United States now demands certified ship manifests to be worked out during the loading of containers for ships proceeding to American ports. This has essentially changed the order of loading in the large ports, for example, in the special administrative region of Hong Kong. International laws adopted by the International Maritime Organization not long ago provide for the defense of the ships, and commercial burdens will fall on the sea transportation industry more severely.

Russia has an interest in the expansion and consolidation of international security for fighting against sea terrorism. Unfortunately, at the present time this fighting is not sufficient. It could be explained by the following factors:

- The lack of a unified system of fighting against international criminal actions both in the world and in the regions;
- Insufficient mutual coordination of state-authorized activities both in the world and in the regions;
- The lack of a legal basis;
- The discrepancy between some countries' national law standards and international agreements signed by them;
- The lack of clear national programs in fighting against international criminal actions on the sea and of those of the punctual distribution of functional duties in that sphere between legal bodies;
- The unsatisfactory equipment of legal bodies and the lack of up-to-date ships in most of the countries, as well as a lack of technical means of communication and identification capable of holding out against the sea underworld;
- The lack of a universally recognized basis of treaty obligations concerning self-defense of the ships and sea navigation's defense from criminal armed attacks;
- An insignificant amount of research on the subject; and
- The lack of a unified international intergovernmental organization engaged in the maintenance of order on the sea and the insufficient participation of the international nongovernmental organizations in this range of activity.

Transportation of Energy Resources and the Turkish Policy in the Black Sea Straits

Vladimir S. Kotliar *

I.

The Black Sea Straits ("the Straits"), which include the straits of Bosporus and Dardanelles and the Marmara Sea, are among the most important and widely used international waterways playing a significant role in the transportation of energy resources. Although statistics of commercial navigation through the Straits show a trend of gradual reduction, the same statistics show that the Straits continue to belong to the category of waterways that are most actively used for international navigation. Official Turkish data on the number of commercial ships passing through the Straits over the last several years can be seen in Table 1.

Table 1

Year	Number of ships
1997	50,942
1998	49,304
1999	47,906
2000	48,079
2001	42,637
2002	47,283
2003	46,939

At the same time, statistics in Table 2 show a growing number of tankers passing through the Straits, especially after the oil pipeline Baku-Novorossiysk, built by the Caspian Pipeline Consortium, began operation.

* Senior Researcher, Diplomatic Academy, Moscow, UN Arbitrator for the Law of the Sea.

Table 2

Year	Number of tankers
1999	5,504
2000	6,093
2001	6,156
2002	7,427
2003	8,097

II.

The present legal regime of the Straits is governed by the Montreux Convention of 1936 that succeeded the Lausanne Treaty of 1923, which was concluded between states that had won World War I and Turkey, as a vanquished state in that war. Whereas the Lausanne Treaty radically limited Turkey's right to control the Straits, conferring this function on the international commission specially established for this purpose, the Montreux Convention ("the Convention") has removed many of those limitations and has restored the sovereignty of Turkey over the Straits under certain conditions. The most important of these conditions is Turkey's obligation to maintain the freedom of navigation of commercial ships by day and night, independent of their flag and cargo, and without any formalities except for a sanitary inspection at the entrance to the Straits.

As for the passage of naval vessels, the Convention took into consideration the importance of the Straits for the Black Sea coastal States as the only exit from the Black Sea, although the Straits already at that stage were fulfilling the function of an international waterway. The Convention empowered Turkey to limit the passage of foreign naval vessels, giving some advantages to the Black Sea coastal States. It should be noted that today no other state in the world that borders straits has similar powers. It should also be pointed out that the government of Turkey, even at the peak of the Cold War, fulfilled in good faith its obligations concerning the passage of naval vessels. This remains Turkey's policy today.

III.

Unfortunately, the same cannot be said about Turkey's policy concerning commercial navigation through the Straits. Over the years Turkey has made several attempts to get rid of the limitations of freedom of its actions in the Straits, which were laid down first in the Lausanne and then in the Montreux Conventions, and to introduce national rules of navigation in the Straits in lieu of international rules. Thus, such attempts have been made in the 1920s and again in 1982, but each time, under the pressure of the USSR and/or other participants of both Conventions, Ankara had to retreat. But in 1994, after the USSR had ceased to exist, Turkey, while not formally refusing to continue implementation of relevant provisions of the Convention, promulgated and put into force its own national Regulations for navigation in the Straits (which, by the way, Ankara had suddenly renamed "the Turkish Straits"). This was done unilaterally, without consulting any of the Convention participants. The substance of the 1994 Regulations and the unilateral manner of their adoption has been judged by the other participants to the Montreux Convention as a clear breach of the Convention.

Thus, these Regulations empowered Turkish authorities to stop commercial navigation through the Straits not only due to *force majeur* (something that other states could at least understand) but also for a number of far less important reasons, such as arranging sports competitions (Article 24). The Regulations also empowered Turkey to be able to introduce limitations on passage of large-capacity tankers or to prohibit their passage altogether (Articles 29, 42 and 52); the Regulations also contained ambiguous formulations allowing for a possibility of police and customs inspections of commercial ships transiting the Straits (Articles 44 and 54).

IV.

Russia and a number of other Convention participants protested those and many other articles of the Regulations, as well as the unilateral manner of their introduction amounting to an obvious violation of the Convention. Since Turkey had refused even to discuss the Regulations, agreeing only to explain their meaning, Russia raised the matter in the International Maritime Organization (IMO). From 1995 to 1998, despite the tough stance of Turkish representatives, sometimes in the form of an ultimatum, the IMO in several decisions supported the Russian position and confirmed the right of any ship to use the Straits in accordance with the Convention.

To a large extent, the IMO took this position because other maritime powers – including the United States and leading European states – had made statements in the IMO to the effect that they shared the Russian view on the inadmissibility of unilateral introduction of national regimes in straits used for international navigation that would limit this international navigation, in violation of the international law of the sea.

It was only after this that Turkey agreed to hold consultations with Russia over its Regulations. The result of those consultations was that in 1998 Turkey revoked the 1994 Regulations, declared its intention to set up a Vessel Traffic Service (VTS) in the Straits to better organize the traffic and to secure safety of navigation through them, and introduced new Regulations, eliminating the most notorious articles of the original Regulations.

Yet Turkey has by no means given up its policy of unilateral actions in the Straits bypassing the Convention. Thus, the new Regulations were adopted by Ankara in a unilateral way and contained loopholes for the Turkish authorities to further limit passage of commercial ships, especially tankers and other large-capacity ships.

V.

However, toward the end of 1998 and in early 1999, the United States and the European Community and NATO states radically changed their position in the IMO on the unilateral actions by Turkey in the Straits, and in effect removed their objections to such actions. This occurred simultaneously with the preparation for and subsequent invasion of Yugoslavia by NATO armed forces. The Clinton administration and European NATO states needed the Turkish support in this war, while the Turkish government needed at least non-resistance from the West to the country's unilateral policy in the Straits, if not consent. And to an even greater extent, the George W. Bush administration and Great Britain needed Turkish support for the unpopular war against Iraq, the preparation for which, as it became known later, started in Washington in late autumn 2001.

On the other hand, the ecological political lobby in the E.C. states was gaining greater influence, especially after the shipwreck of the tanker Prestige off the Spanish coast, which led to the introduction of tougher ecological E.C. legislation.

Turkey saw its chance, and in 2002 took a new unilateral step adopting the Instruction on the application of the 1998 Regulations, again in a unilateral manner. Among other things, the Instruction has toughened the national regulations for navigation in the Straits in cases of passage of tankers and other large-capacity ships 250-300 meters long and with tonnage of 50,000 tons or more by prohibiting any traffic in the opposite direction during their passage. This time, Russia's protests in the IMO against those measures were not supported either by the United States, or by the E.C. states.

VI.

Ankara tries to justify its unilateral actions with an argument that Turkey, which has sovereignty over both shores of the Straits, has a duty to take measures to strengthen both the safety of navigation and the security in the Straits, and to protect its cultural heritage and the environment as well as human life and property in the Straits zone.

Turkey's concern with the need to strengthen the safety of navigation in the Straits is understood and supported not only in the United States and the E.C. states, but also in Russia. But the point is that Turkey has been introducing measures to this effect in a unilateral manner, which does not correspond to the Convention, and the concern with the safety of navigation is not the only motive for Turkish actions. It was clear that the policy of limiting the international navigation in the Straits through substituting the Turkish national jurisdiction over the Straits for the Convention regime had begun to be especially actively conducted by Turkey only with the start of construction of the oil pipeline Baku-Novorossiysk and with the launch of the competing oil pipeline project Baku-Tbilisi-Ceykhan, which is supported by the United States and which promises considerable income for Turkey. However, while the Baku-Novorossiysk pipeline has been operating successfully for over three years now, it has become doubtful that there will be enough oil to fill the Baku-Tbilisi-Ceykhan pipeline. In other words, Turkish policy in the Straits seems to be dictated to a large extent by the Turkish aspiration to limit the

Russian oil export through the Straits and thus to secure more effective use of the Baku-Tbilisi-Ceykhan oil pipeline.

This latter circumstance gives rise to a legitimate concern in Russia, which President of the Russian Federation Vladimir Putin and former Russian Prime Minister Mikhail Kasyanov did not hide in their public statements in Eisk and Novorossiysk, respectively, in 2003. In March 2003, Mr. Igor Ivanov, Russian Foreign Minister at the time, in a letter to his Turkish colleague, stressed the importance of the Straits for both countries. He pointed out prospects for their fruitful cooperation, and expressed his hope that Turkey would take appropriate steps to secure the non-interrupted functioning of the Straits as an international transit way and avoid obstacles that might jeopardize constructive cooperation between Russia and Turkey.

Another Turkish motive may be purely political – by the introduction of stricter national control over the Straits, Turkey obviously hopes to enhance its political weight in its relations both with Moscow and the United States. At the same time, having construction finished and the Baku-Tbilisi-Ceykhan oil pipeline functioning would strengthen the political influence of Ankara in all those Caucasian and Central-Asian countries whose budget will depend on its optimal use.

VII.

Russia, for its part, is quite aware that there are good reasons behind the Turkish concern with the need for safer navigation in the Straits and for additional measures to be taken to achieve this – after all, since the Convention was adopted in 1936, the number of ships passing through the Straits annually has grown 10 times. However, Russia disagrees with the Turkish arguments that all measures taken by Turkey in the Straits have been motivated by its duty to secure the main principle of the Convention, i.e., the freedom of navigation (since in the event of a major shipwreck the navigation in the Straits may be interrupted for a long time), therefore no one should object to them. Russia also disagrees with the unilateral manner in which Turkey has been adopting its measures, without consulting participants to the Convention, bilaterally or through the IMO, and thus its national Regulations in practice substitute for the Convention.

The main Russian objection is that Turkey has no legal right to unilaterally pass judgment on the need for limitation of freedom of navigation in the Straits and the degree to which it is needed. It was on the condition of the maintenance of that very freedom of navigation in the Straits that the Montreux Convention has returned sovereignty over the Sraits to Turkey. Sovereignty does not absolve states bordering straits of their duty to abide by the international law of the sea (in this case, by the Convention). That is why the Foreign Ministry of Turkey's reference to the Turkish sovereignty over the Straits to justify the unilateral nature of its regulatory measures in the Straits simply does not work. This is the clear message of Article 34 of the UN Convention on the Law of the Sea. This is not the proper time or place to go into a detailed analysis of the Turkish position, but generally speaking all legal arguments by the Turkish side in support of the Turkish "right" to quietly substitute national Regulations for the Montreux Convention, or attempts to present notions of "freedom of navigation" and "safety of navigation" as incompatible (although, given good will, they are fully compatible), are beneath criticism. Besides, both the Regulations of 1998 and the Instruction of 2002 contain a number of inadequate provisions that create quite unnecessary risks of shipwreck and fail to enhance safety of navigation. As Russia informed the IMO in March 2003, Turkey's adoption of this Instruction has, compared with 2001, led to a 400 percent increase in ship delays at the entrances to the Straits. When dozens of large ships stay for hours in dynamic drift in relatively small sea areas around the entrances to the Straits, it only creates an additional risk for their safety. The VTS system constructed for Turkey by the U.S. Lockheed-Martin, which began operating in the Straits in 2003, up to now has been used mainly to gather statistical data on passage of ships, but it has not yet led to a more effective use of the Straits.

VIII.

A question also arises of whether people in positions of responsibility in Washington and Brussels have given thought to the long-term consequences of their consent to the virtual replacement of the Montreux Convention with the Turkish national Regulations and other Turkish decisions affecting the navigation through the Straits adopted by Ankara in a unilateral way, without prior consultations with participants to the Convention or member States of the IMO.

It was exactly this substitution of national rules of passage through straits used for international navigation for international regime of such straits that the United States has hitherto opposed most actively. Efforts have been undertaken by the United States, Russia and other major maritime powers to prevent a unilateral introduction, without consultations within the IMO framework, of national rules of navigation in the Singapore and the Malacca Straits as well as through archipelagic corridors in the Indonesian archipelago. But if Turkey, with U.S. consent, could arrange such substitution in the Straits without consultations with anybody, why can not dozens of other states bordering straits attempt to repeat this experiment?

As for the E.C. countries, Russia is currently the second source of oil supply to the European market after the Middle East. Of 144 million tons of oil delivered by Russia to Europe in 2002, almost 50 million tons, or one third, came from the Black Sea oil terminals in Novorossiysk, Odessa and Tuapse. Is Europe prepared to let Turkey use the dependence of the E.C. states on those deliveries via the Straits and, by changing unilaterally the rules of navigation in the Straits, to play its own games, for instance, in such matters as Turkey's E.C. membership or recognition of the Northern Cyprus?

IX.

Today the Straits continue to be of great strategic and economic importance to Russia. But Russia is far from aspiring to turn the Straits into a gigantic oil pipeline as Turkish governmental officials sometimes state. Russia, like Turkey, does not want to see tanker shipwrecks in the Straits with fatalities and large-scale pollution of the marine environment. On June 16, 2004, Mr. Sergei Lavrov, Foreign Minister of the Russian Federation, stressed in an interview for the Turkish press that despite the shortage of its oil export facilities, Russia was not planning any substantial increase of oil deliveries via the Black Sea in the near future.

It is also clear that a growing number of tanker passages through the Straits might call for additional measures to secure the safety of navigation. But every such measure will limit that very freedom of navigation in the Straits that Turkey is treaty bound to maintain. This is why Turkey has no right to adopt these measures in a unilateral way, thus willfully limiting or changing its international obligations.

In the same interview Mr. Lavrov, as if responding to Turkish statements to the effect that the passage capacity of the Straits would have been fully exhausted, pointed to the need "to maximize an effective use of the main routes of oil transportation, in full conformity with the international law." He said, "Russia had been conducting active consultations with Turkish colleagues on the ways to increase the effectiveness of navigation in the Straits taking into consideration the fact that there remained reserves to increase their passage capacity. By the way, it is the view not only of our specialists but also of experts from reputed international institutes and organizations," he said, probably referring to, among others, a similar conclusion of a study made several years ago by the experts of Lloyds-Register, a British firm well-known in international maritime circles.

Indeed, the Russian-Turkish consultations on the Straits have been taking place on a regular basis for several years now, and there is no doubt that this useful practice should be continued. Unfortunately, today they are held only for the purpose of exchanging views on those Turkish Regulations that have already begun operation in a unilateral way. However, these consultations would contribute to a far larger extent to more effective and safe navigation in the Straits if Turkey also used them to discuss its new regulations at the planning stage. It is this joint preparatory work on such regulations by Turkey and other participants to the Convention, with their subsequent adoption by Turkey, that would be the most effective approach and the only legitimate one, from the viewpoint of international law. Only this approach could help to preempt misunderstandings and conflicts over new regulations, the introduction of which would then present no surprises. Similarly, this approach could help to prevent an escalation of cloning in the other straits of the World Ocean, in circumvention of the UN Convention of the Law of the Sea, of the Turkish precedent of introduction of national regime of navigation in the Black Sea Straits in circumvention of the Montreux Convention of 1936.

Legal Aspects of Russian Perspectives on Continental Shelf Issues in the Arctic Areas

Alexander S. Skaridov[*] *and Mariya A. Skaridova*[†]

On December 20, 2001, the Russian Federation made a submission through the Secretary-General to the Commission on the Limits of the Continental Shelf, pursuant to article 76, paragraph 8, of the 1982 United Nations Convention on the Law of the Sea. The submission contains data and information on the proposed outer limits of the continental shelf of the Russian Federation beyond 200 nautical miles from the baselines from which the breadth of the territorial sea is measured (often referred to as an extended continental shelf). The Convention entered into force for the Russian Federation on April 11, 1997.[1]

In accordance with rule 49 of the Rules of Procedure of the Commission on the Limits of the Continental Shelf (CLCS/3/Rev.3 and Corr. 1),[2] which requires that the proposed outer limits of the extended continental shelf pursuant to the submission be made public, a note verbal was circulated to all member States of the United Nations, including the States Parties to the Convention. The note verbal contains information regarding the outer limits of the Russian continental shelf in the Arctic and Pacific Oceans. It includes geographical coordinates of points delineating the proposed outer limits, as well as illustrative maps.[3]

The agenda for the 10th session of the Commission, which was held in New York from March 25 to April 12, 2002, included consideration of the Russian Federation's submission. On June 27, 2002, after an examination of the scientific data and other material included in the submission, the Commission adopted its recommendations, which were sent both to the Russian Federation and to the Secretary-General.

The Commission presented its recommendations for the four areas contained in the submission relating to the continental shelf extending beyond 200 nautical miles: the Barents Sea, the Bering Sea, the Sea of Okhotsk and the Central Arctic Ocean. The Commission requested that the Russian Federation revise the submission in certain areas. The 12th

[*] Professor of Russian State Humanitarian University Faculty of Law.
[†] Scientific Fellow of the Russian Institute for Ocean Law Studies (AHE).

session of the Commission on the Limits of the Continental Shelf took place April 28 through May 2, 2003.

The continental shelf is defined in the Convention as the seabed and subsoil of submarine areas that, because of their geological characteristics, are considered to be the natural prolongation of the continental or land mass beneath the oceans or seas to the outer edge of the continental margin, or to a distance of 200 nautical miles from the baselines from which the breadth of the territorial sea is measured. The continental margin consists of the shelf, the slope and the rise. It does not include the deep ocean floor with its oceanic ridges or the subsoil thereof. The Convention gives coastal States sovereign rights to explore and exploit such resources, which are thought to be extremely valuable.[4]

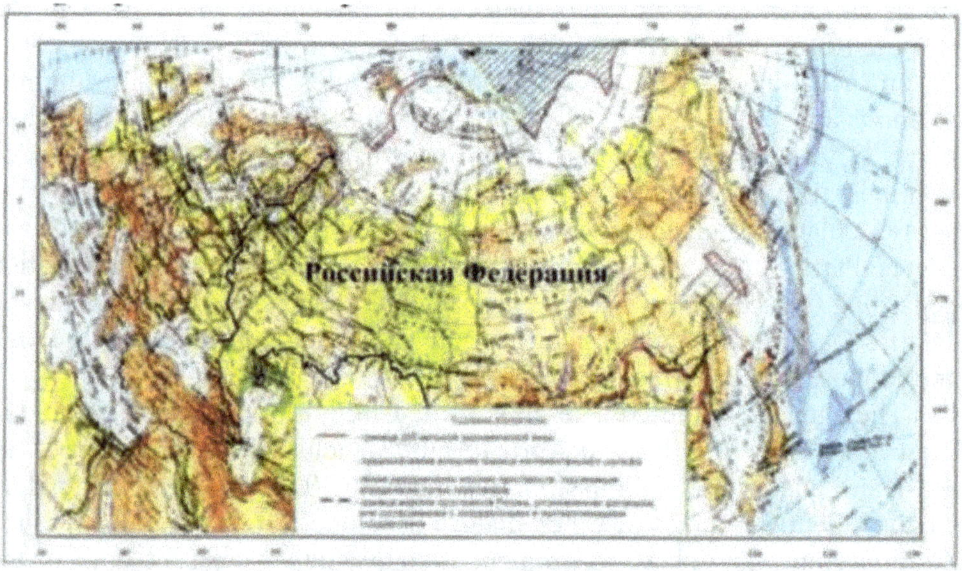

Map 1: Illustration of limits of the economic zone and the continental shelf of Russia. Legend in English:

1. Limits of the 200-nautical-mile economic zone
2. Anticipated outer limit of the continental shelf
3. Line of delimitation of maritime zones, subject to determination through negotiations
4. Limit of the maritime zones of the Russian Federation established by agreements or understandings with the adjacent or opposite States

Russia has the most extensive continental margin in the Arctic, and this margin is the region of special scientific, political, economic and defense interest.

According to Federal Law N 30-FZ of February 26, 1998, Russia has ratified the 1982 UN Convention on Law of the Sea. This requires the definition and validation of the Continental Shelf Outer Boundary (CSOB) in the Arctic outside the limits of the 200-mile zone in the juridical context of the Convention.

According to the Convention, bathymetric, geomorphological and geological criteria are to be used for the establishment of the CSOB outside of the 200-mile zone.

The main purpose of this paper is to demonstrate the Russian position on determining the outer limit of the continental shelf of the Russian Federation. I am not going to cover the case of the outer limit of Russia's continental shelf in the Bering and Okhotsk seas, but will concentrate on the Arctic basin.

As Professor Carrera (Hurtado, Galo) wrote in one of his papers, the full implementation of article 76 of the United Nations Convention on the Law of the Sea (UNCLOS) poses one of the most significant scientific challenges in the fields of geodesy, geology, geophysics and hydrography during the next decade. These challenges stem from the collection, compilation, and processing of vast amounts of marine data to be presented to the Commission on the Limits of the Continental Shelf as evidence in support of continental shelf claims to national jurisdiction extended beyond 200 M.

The geological validation of the position of CSOB is based on the recognition of the geological integrity of oceanic structures. The boundaries of these structures, determined by bathymetric, geomorphological and geological criteria, may contradict juridically established shelf boundaries. Therefore, the comprehensive investigation of the structure and nature of the earth's crust (continental or oceanic) becomes one of the most important geological problems. Among legal problems related with the subject of this presentation, I will underline four aspects that will influence the delimitation of Arctic spaces:

Map 2

MAP 2: Area of the continental shelf of Russia in the Arctic Ocean beyond the 200-nautical-mile zone.
Legend in English: Outer limits of the Continental Shelf determined according to various criteria:

1. Provisional line of the delimitation of the continental shelf of the Russian Federation with neighboring States; subject to more precise determination through negotiations;
2. 200-nautical-mile zone from the baselines;
3. Line showing the thickness of the sedimentary rocks equal to 1 percent of the distance from the foot of the continental slope;
4. 60-nautical-mile zone from the foot of the continental slope;
5. Points of the outer limits of the continental shelf determined according to various criteria, their numbers;
6. Numbers of segments of the outer limits of the continental shelf determined according to various criteria;
7. Baselines;
8. Basepoints;
9. 200-nautical-mile zone from the baselines; and
10. The area of the continental shelf of the Russian Federation in the Arctic Ocean beyond 200-nautical-mile zone.

Scale: 1:10,000,000 [Note: Not to scale - for the purpose of this paper, the map has been reduced from A1 to letter size format.]

(1) Accuracy of Straight Baseline Determination;
(2) Absence of Internationally Accepted Map Showing 2500m Contour for the Arctic Spaces;
(3) Territorial Dispute; and
(4) Russian Scientific Data Used to Support Russian Claims on the Continental Shelf.

THE ACCURACY OF STRAIGHT BASELINE DETERMINATION

In December 2001, the U.S. Embassy in Moscow sent a note to the Ministry of Foreign Affairs of Russia (MFA N 120) on Russia's submission of documents on outer limits of the continental shelf. This note stated that the U.S. government considered that the Russian government should detain its application. One of the reasons was that some baselines demarked by Russia, which were set by the USSR in 1985, did not satisfy UNCLOS requirements.

Articles 7, 9, 10 and 47 of the United Nations Convention on the Law of the Sea define the circumstances under which coastal States can construct straight, closing and archipelagic baselines. However, the subject Articles do not specify, or require coastal States to specify, the geometric or technical properties of these lines. As a consequence, coastal States are free to apply any interpretation they choose in the technical definition of straight baselines, whether these are loxodromes, arcs of great circles, arcs of small circles, grid lines, normal sections from either terminal of a segment, or geodesics. Coastal States have occasionally defined the technical properties of straight baselines through promulgation in national maritime legislation or in international maritime boundary delimitation treaties, but this has been very much the exception rather than the rule.

The absence of a definition of the geometric properties of straight, closing or archipelagic baselines from national maritime legislation is undoubtedly due to the non-explicit nature of the wording of UNCLOS Articles 7, 9, 10 and 47. It could be argued that this situation has not been improved by some coastal States that simply follow the precedent set by others in the wording of their own legislation. However, coastal States can take advantage of this situation as it enables the choice of a particular line type that can be used to gain a territorial advantage. For example, under certain circumstances, the use of loxodromes rather than geodesics can sometimes lead to a significant maritime territorial gain, especially where long lines are involved. The converse case can also occur, with a geodesic offering an advantage over the use of a loxodrome.

However, Article 76 (7) states the following: "The coastal State shall delineate the outer limits of its continental shelf, where that shelf extends beyond 200 nautical miles from the baselines from which the breadth of the territorial sea is measured, by straight lines not exceeding 60 nautical miles in length."

Russia responded to the U.S. government's arguments (statements made by the Russian delegation at the meeting of Arctic States on November 26 and 27, 2001, in New York) that straight baselines for setting a 200-mile EEZ were not used in Russia's application, only those baselines that were approved by the Council of Minister's decrees in 1984 and 1985 (with a few specifications).

Furthermore, applying the baseline rules in these ice-covered areas is an exceedingly complex task. We are dealing with the permanently frozen ice shelf, and the baselines in some places could be drawn from the edge of permanent ice which will "deviate" from the coastline. This could freeze discussions for many years. Nine months is the longest amount of time the polar seas are covered with ice, a period in which regular vessels cannot pass through the seas. It is almost impossible to determine where the land ends and sea ice surface begins.

Absence of an Internationally Accepted Map Showing 2,500m Contour for Arctic Spaces

The fixed points comprising the line of the outer limits of the continental shelf on the seabed, drawn in accordance with paragraph 4 (a)(i) and (ii), either shall not exceed 350 nautical miles from the baselines from which the breadth of the territorial sea is measured or shall not exceed 100 nautical miles from the 2,500 meter isobaths, which is a line connecting the depth of 2,500 meters.

As it known, the 2,500m contour plus 100 nautical miles line is one of two possible outer constraints to continental shelf limits under Article 76 and will consequently form the outer limit for some coastal States in some areas. Consequently, its location may be of extreme importance to some limit determinations. Article 76 defines the 2,500m contour as "a line connecting the depth of 2,500 metres", a definition that emphasizes that the 2,500m contour is a linear feature comprising depth and connection. Worldwide, few areas containing the 2,500m contour have been systematically mapped, with most continental slope areas being covered by assemblages of individual single-beam tracks collected over many years.

The 2,500m contour is to be used as the line from which a constraining line can be constructed, at a distance of 100 nautical miles seaward from it. This 2,500m plus 100 nautical miles line is one of two lines that must be combined to form a constraint on the outer limit of the area that a state can claim. The other constraint line is one "350 nautical miles from the baselines from which the breadth of the territorial sea is measured." The two constraints are blended together by choosing sections of whichever is most seaward. Potentially, any misplacement of the 2,500m contour can substantially affect the area that could be claimed by a state beyond 350 nautical miles. Consequently, its mapping deserves careful consideration.

The internationally accepted map showing the 2,500m contour for the entire earth is GEBCO (IHO, IOC and CHS, 1984; Jones, 1997). The 2,500m contour as depicted by this map does not include the Arctic. In shallow waters, particularly those used for navigation, there is a necessarily strong interest in determining all errors in depth measurement.[5]

In the Russian submission the bathymetric and morphological criteria are based on the identification of the 2,500 meter isobath and the location of the continental rise, respectively. Russian investigations made it possible to accurately determine the positions of 2,500 meter isobath and continental rise and to plot them on the basis of original bathymetric survey with the accuracy Moi <600m and Mzi < 0.5%.

Territorial Dispute

Despite the USSR and Norway agreement on marine borders, signed on February 15, 1957, and the bilateral negotiations of 1970, the issue of delimitation of economic zones and continental shelf in Barents Sea was not solved. In 1975, Norway proposed to make shelf delimitations along the line lying east of the polar area of the USSR This way Norway claimed 155,000 km^2 of Russian marine space. It also claims 10,000 km^2 as its economic zone, while straightening polar areas of Russia to the north off Spitzbergen's border. In this case Russia loses a water area of 25,000 km^2. To adjust these problems, on October 15, 1976, the USSR and Norway signed an Agreement on Mutual Relations in Fishery, and on

January 11, 1978, a Protocol on temporary regulations for the fishery in the so-called "gray zone" of the Barents Sea was signed.

The U.S.-Russian agreement of 1999 did not define the status of the Wrangel and Herald Islands and De Long Islands (Bennett, Henrietta, and Jeannette Islands) in the Arctic. Because no one agreement defines the status of the islands, the U.S. claims sovereignty over the territories. The U.S. position is that the question of ultimate ownership is not prejudiced by the treaty and remains open, but a legal situation regarding these islands will remain one of conflicting claims under international law.

Furthermore, because of reasons mentioned previously, applying the baseline rules in these ice-covered areas is an exceedingly complex task as it deals with the permanently frozen ice shelf and the baselines in some places could be drawn from the edge of permanent ice. The Arctic Ocean is completely encircled by the continental shelves of five coastal States, and simple geometric considerations suggest that implementing the provisions of Article 76 of the Law of the Sea could easily lead to overlapping outer limit claims. Forming hypothetical outer limits of economic zones and continental shelves, settled according to UNCLOS (see maps 3 and 4), shows the difficulty of their delimitation and the presence of "gray zones." This can make a delimitation issue a long and complicated procedure that lasts years.

Disputes may well be inevitable as each coastal State attempts to maximize the area of its juridical continental shelf; however, a significant cause of disagreement among the affected states could be eliminated through the adoption of common descriptions of bathymetry and sediment thickness. The development of a standard bathymetric database is now in the works, and the creation of a complementary database of sediment thickness is a topic of discussion.

Russian Scientific Data Used to Support Russian Claims on the Continental Shelf

The Commission on the Continental Shelf's denial of the Russian submission would indicate the Commission was not satisfied with Russian arguments, which means the Commission disposed with precise scientific data, the data presented by Russia. This seems to be questionable. Several Russian governmental institutions are involved in gathering scientific data in the Arctic.[6]

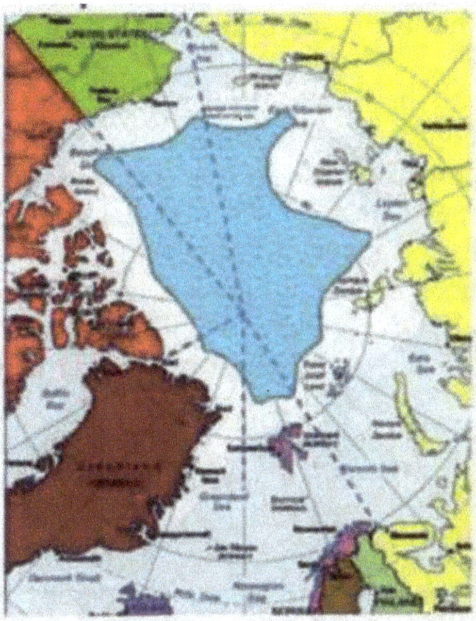

Map 3

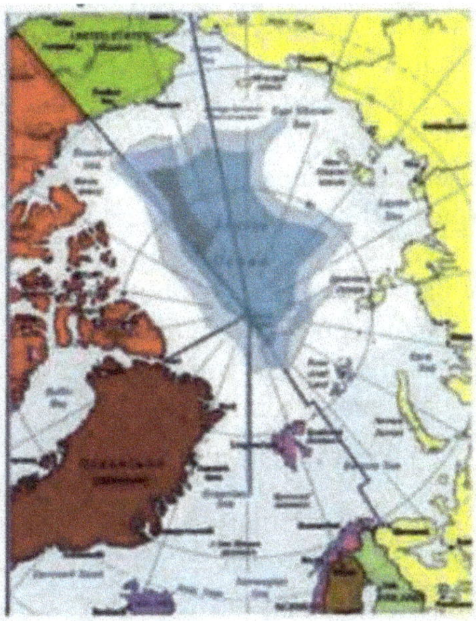

Map 4

Establishing continental shelf limits in the Arctic Ocean beyond 200 nautical miles in accordance with Article 76 of UNCLOS appears a highly complex procedure requiring a thorough analysis of a wide variety of information, especially evidence reaffirming that the claimed area belongs to the continental margin in the sense perceived by article 76.

Such evidence is, in turn, to a large extent based on scientific interpretations of a large spectrum of observation data acquired by different research disciplines: hydrography, geomorphology, geophysics, and geology.

For many years, Naval activities in the Arctic were connected with the military-strategic interests in this region of the world ocean, based on the creation of effective infrastructure for the naval strategic nuclear forces and multi-mission forces along the Arctic sea frontiers of Russia. The unique features of Arctic natural conditions and their specific influence on the employment of the naval forces and facilities stimulated the active participation of the Navy in solving the fundamental problems of the physics of the ocean and atmosphere and resulted in long-term experience with organizing the activities of complicated organizational and technical systems under extreme Arctic conditions. One of the most important scientific and practical results that arose from these activities was the establishment of an information-technological infrastructure of navigation and hydrographic support that is unique in its capabilities and resources.

For several decades when the Arctic basin was being developed, the Russian Navy carried out a colossal scope of research in diverse fields of knowledge, enlisting the services of advanced scientific and industrial organizations. Eighty percent of the Arctic Ocean and its marginal seas are covered by oceanographic, gravimetric and magnetic surveys, as well as by bathymetric and sea bottom soils surveys, except for some areas of the Canadian sector. The volume of these data exceeds that of all the rest near Arctic states taken together; the data are sure to be of interest for solving fundamental scientific problems.

Many of the resources that were inaccessible for civil and foreign users have been declassified and are coming into scientific and commercial circulation.

Such knowledge provides users with high-quality cartographic materials, including the electronic navigational charts prepared in compliance with international standards and covering the greatest part of the Arctic water area. And the prospects of creating the automated distribution system of digital cartographic information open it up to new opportunities of unrestricted access for any interested organizations and mariners in compliance with user service discipline in force.

Russian scientists have characterized the morphology and deep structure of the Arctic Ocean based on airborne magnetic and gravity surveys, regional refraction, reflection and deep seismic surveys. These data were obtained over a period of more than 30 years, mostly from drifting stations such as "Severny Polyus" (North Pole) and from high latitude expeditions such as "Sever." Some geological and geomorphologic results indicate that continental crust is not limited by the continental slope but may extend far into the Arctic Ocean. The processing of geophysical data using up-to-date software and hardware made it possible to study in detail the structure of the sedimentary cover, the lower crust and the upper mantle. The following new data relevant to determining the CSOB were obtained:

1. The thickness of the sedimentary cover on the Lomonosov Ridge is 5.0 - 5.5 km. The seismic velocities of the upper (6.0 - 6.6 km/s) and lower (6.8 - 7.5 km/s) crust imply a continental pattern of the ridge.
2. Total crust thickness in the Podvodnikov Basin is 20-25 km. Along with the thick sedimentary cover (10 km), low velocities of the lower

crust, and the mantle structure. The new findings imply a continental nature for the Podvodnikov Basin.
3. In the Makarov Basin, a thick (up to 7 km) sedimentary cover at a depth of 12 km rests on a thin (up to 2-3 km) oceanic layer or directly on the mantle. Based on these we presume that the crust of the Makarov Basin is very similar to the oceanic type.
4. The data obtained so far for the Alfa-Mendeleev Ridge does not allow an unambiguous interpretation of the nature of its crust.[7]

The current concepts of the deep structure of the Arctic region, data on the thickness and structure of the sedimentary cover, as well as geomorphologic and bathymetric data were used to compile a project (see Map 2) showing the present Russian viewpoint on the position of the continental shelf outer boundary in the Arctic Basin.

Russia conducted the program of Arctic science cruises aboard nuclear-powered submarines from the early 1960s, providing the only regular, reliable access to the deep Arctic Ocean basin for science. Submarines operating efficiently in the Arctic, below the permanent floating pack ice, conducted the first systematic surveys of Arctic bathymetry and water composition. These cruises yielded new data on the structure of the ocean basin, and the distribution of heat, salinity and composition of the water. Peaceful use of these submarines has provided an extraordinary opportunity to explore the Arctic Ocean basin.

The U.S. Navy has also carried out the geophysical program from submarines focusing on the Gakkel and Lomonosov Ridges and the Chukchi Borderland, and also in operational areas defined by the exclusion of all non-U.S. EEZs and a few shallow shelf areas. In 1999, for the first time, the USS *Hawkbill* collected data in the EEZ of Norway, along the Yermak plateau, operating in these waters at the invitation of the Norwegian government.[8]

The TRANSARCTIC 2000 and ARCTIC 2003 expeditions, organized by the Russian Ministry of Defense and the Minister of Natural Resources, proved the 6.5 million square mile continent seabed in East-Siberian and Chukcha seas was not volcanic in nature.

Let me consider some of the criteria contained in the Convention for the delineation of the outer limit of the continental shelf.

Article 76 establishes two criteria for establishing the outer edge of the submarine continental margin: the first criterion is based on the thickness of sedimentary cover (paragraph 4 (a) (i)), i.e., the thickness of sedimentary cover within the outer limit of the continental shelf must be at least 1 percent of the distance from the foot of the continental slope (sediment thickness formula).

The second criterion, which is based on distance, is set forth in paragraph 4 (a) (ii); the outer limit of the submarine continental margin is determined as 60 nautical miles from the foot of the continental slope (distance formula).

Article 76, paragraph 5, sets the following restrictions on determining the position of the outer limit of the continental shelf:

(a) 350 nautical miles from the baselines; and
(b) 100 nautical miles from the 2,500 isobath.

The coastal State has the right to use the most advantageous combination of constraint lines and formula lines to extend its continental shelf.

The Russian approach to the delimitation of the submarine continental margin in terms of article 76 is based on its own analysis of the earth's crust under the Arctic Ocean, prepared in accordance with the findings of geological and geophysical research.[9]

Source materials from the 1960 to 1990 period were used in constructing the foot of the continental shelf ("FCS") and the 2,500 m isobaths. During that period, almost the entire area of the marine basin was surveyed by high-precision systematic depth soundings with a density of measurements ranging between five and 15 km. A total of 21,120 depth points were measured by echo-sounder, and 17,426 depth points were measured by seismic soundings. Depth soundings from submarines covered 90,716 linear kilometers. The precision of the depth measurements remained within 0.5 percent of the measured depths, and were corrected for sound velocity in water using the findings of observations from hydrological stations which were carried out at depth measurement points, as well as tabulated data for the correction of depth values measured by echo-sounder.

Prior to 1969, the positioning of the depth measuring stations was determined by astronomic-geodetic means with average accuracy of about 1,000m. Subsequent satellite navigation systems were used for this purpose, and the accuracy of determining the position of the stations moved into the range of 150 to 400 meters.

Over large areas, the bottom topography was surveyed by means of brief landings of one or more helicopters or aircraft delivering research teams and seismic equipment to small ice floes. The large-scale survey covered an area of 4.1 million sq. km, or 80 percent of the entire area of the Arctic basin.

I believe the Soviet Navy was among the first navies prior to 1969 to start realizing a scientific program of Arctic exploration. During the same period, depth soundings from submarines covered approximately 91,000 km of linear kilometers with a mean accuracy of depth point positioning within 1.0 to 1.5 miles; readings were taken by NEL-6 "Mologa" echo-sounders.

On the basis of the findings of the Russian hydrographic surveys, a bathymetric map of the Arctic basin was prepared with a scale of 1:5,000,000, and 200m contour intervals. An analysis of the bathymetric data and mapping of the bottom topography was carried out using morphological constructions, allowing a detailed representation of the various features of the topography. All the positive and negative forms of the topography were mapped in detail, showing their close interrelationship within the Arctic basin.

The findings of the surveys and mapping of the topography made it possible to establish the position of the basic criteria of article 76 of the Convention 2,500m isobaths on the continental margin. Within each composite profile, isobaths were spaced at 200m intervals on a strip 40 to 50 km wide, and depth values were shown at the points of their intersection with the profile. When the profile was converted into digital format, it was split into equal 2,500m intervals and the depth values at these fixed points were determined by linear interpolation.

Taking into account the mean accuracy of the initial bathymetric data used in the calculation (of about 600m), the mean accuracy of the FCS points was close to that value. However, since some measurements of positioning points had errors of up to 3,000m, the accuracy of the FCS points was in the same error range, which is indicated in the submission.

The location of the exclusive economic zone of the Russian Federation (the 200-mile zone) was calculated by using specially developed software.

The position of the first constraint line, the 350-mile zone from the baselines of the Russian Federation, which was determined by forward position computation, is shown on Map 4.

The position of the 2,500m isobaths was calculated on the same profiles that were used for determining the FCS. The coordinates of the points of the 2,500m isobaths were calculated by the method of linear interpolation of coordinates and depth values relating to adjacent points of the profile, on the basis of the formulae specified in the submission. The accuracy of the positioning of the 2,500m isobaths corresponds to that of the determination of the FCS.

In accordance with article 76, paragraph 5 of the Convention, the fixed points comprising the line of the outer limits of the continental shelf shall not exceed 350 nautical miles from the baselines of the coastal State (constraint line I), or 100 nautical miles from the 2,500m isobaths (constraint line II). The outer limit of the continental shelf of the Russian Federation does not extend beyond constraint line I in the Eurasian basin and constraint line II in the Amerasian basin.

All of the geological, geophysical and bathymetric data made it possible to determine the position of the outer limit of the continental shelf of the Russian Federation in the Arctic Ocean in accordance with article 76 of the Convention.

Concepts concerning the structure and type of the earth's crust underlying Lomonosov ridge, Mendeleev rise and Alpha ridge are of fundamental importance for the Russian submission. To this end, all available geological and geophysical data on the Arctic basin were analyzed, and special additional field investigations were carried out.

In order to determine the structure and nature of the crust of the basic morphostructures of the Amerasian basin, special field investigations were undertaken in 1989 through 1992 and in 2000. Investigations were carried out using the deep seismic sounding and seismic reflection methods, and measurements of potential fields. The investigations on Mendeleev rise by the "Transarctica-2000" expedition were supplemented by geological studies (bottom sampling using gravity corer, grabs and dredges). Ice-resistant research vessels of the "Akademik Fedorov" type with on-board aircraft (MI-8 helicopters) were used for complex field investigations.

The conclusion regarding the continental nature of the earth's crust was also made on the basis of the interpretation of seismic data (deep seismic sounding and seismic reflection sounding) obtained along the geotraverse across the Mendeleev rise during the "Transarctic-2000" expedition and the NP-26 ice-drifting station. The earth's crust is up to 32 km thick and contains typical components of the continental crust (sedimentary layer up to 5 km thick, upper crust up to 6-8 km thick, with velocity reversal in the lower part, and lower crust up to 20 km thick).

The final outcome of the delineation of the outer limit of the continental shelf of the Russian Federation in the Arctic Ocean is the area of the continental shelf situated beyond the limits of the 200-mile zone over which the Russian Federation has jurisdiction.

On June 27, 2002, after an examination of the scientific data and other material included in the submission by the Russian Federation, the Commission adopted its recommendations, which were sent both to the Russian Federation and to the Secretary-General.

The Commission requested that the Russian Federation revise the submission in certain areas. After examining the submission, the Commission in 2003 rejected the Russian submission.

CONCLUSIONS

The Arctic Ocean is almost completely surrounded by the land masses of five coastal States: Canada, Greenland (Denmark), Norway, Russia, and the United States. Each of these states appears to be in a position to develop an extended outer limit of its juridical continental shelf on the basis of a "natural prolongation of its land territory" projecting northward into the Arctic Ocean.

The Commission based its decision on the arguments that in some Arctic areas, the development of accurate and credible outer limits is hampered in several areas, as the information needed to satisfy the criteria of Article 76, namely water depth and sediment thickness, is absent. Moreover, basic geometric considerations suggest the possibility that states' claims will not overlap the 200 nautical mile limit.

Some experts think published portrayals of the seabed of the Arctic Ocean are not precise, and in many places charted depths are fragmented. In other words, the main reason for this situation has been a lack of information needed to construct reliable and detailed charts: certain regions remain poorly mapped on account of difficult operating conditions. It is obvious that knowledge is limited and will be limited in the future. But few countries explore the Arctic on a regular basis, and few countries are able to use nuclear powered submarines for this purpose. On the other hand, Russia, as well as other Arctic and non-Arctic states, will not have precise scientific data on this matter, which means that the

Commission will not have valid grounds for developing continental shelf claims beyond their 200 nautical mile limits in the Arctic Ocean.

Current situations yield three obvious conclusions:

(1) Even a cursory examination of existing information reveals at least seven natural prolongations extending from the continental margins of Canada, Greenland (Denmark), Norway, Russia, and the United States. In clockwise order, the prolongations consist of: Chukchi Cap, Mendeleev Ridge, Lomonosov Ridge (both ends), Yermak Plateau, Morris Jesup Plateau and Alpha Ridge.
(2) It is also easy to see that some of these features could affect the continental shelf configurations of more than one state, raising the possibility, if not the likelihood, of overlapping claims between neighboring states.
(3) The Arctic resources exploitation is an ongoing process that will influence the solution on the legal status of the Arctic seabed.

It is estimated that the hydrocarbon reserves in the Russian Arctic are about 113 billion tons. Today the most interesting natural resources fields are situated in areas called "transition zones." These are belts of shallow water directly adjacent to a shore line of a marine basin, including banks, river deltas, shallow-water reefs, broad tidal zones, littoral zones, and shallow-water coastal sectors, where water depth is generally less than 10 to 15 meters. Such fields as Kharasavey in Yamal and Medyn in the Pechora Lowland are divided by the shore line into almost equal parts. The zone of the Russian Shelf, with shallow water depths of zero to 20 meters, extends over an area of 570,000 square kilometers, and the Arctic Shelf is assessed to contain up to 25 percent of the total Arctic offshore resources.

In the area west of the Arctic Shelf are two large oil-gas provinces, the transition zones of which have high potential. One of these is the broad shallow-water area adjacent to the Pechora Lowland and Kolguyev Island. Another broad, favorable belt of shallow water is in the Kara Sea. This is a direct continuation of the West Siberian province. Favorable shallow-water zones are known to exist in the eastern Arctic Sea, but they have received little study. The Ust'-Khatanga and Ust'-Lena shallow-water areas are the most favorable. The latter is located on a continuation of the Mid-Atlantic Ridge, and geological conditions are similar to those of California. Unfortunately, the weather conditions are a little bit different.

Under the forecasts of the Ministry of Natural Resources, the prospective initial reserves of hydrocarbons on the continental shelf around Chukotka and on land are valued as follows (in conditional propellant):

East/Siberian Sea	5.6 billion tons
Chukotka	3.3 billion tons
Bering Sea	1.1 billion tons (1.5 billion on the continental section of the peninsula, based on geological data)

On the Chukotka shelf only the Bering Sea is expected to have a discovery of four oilfields with reserves of between 100 and 300 million tons. It is also expected to have smaller oilfields, including nine with reserves of between 30 and 100 million tons, three gas fields of between 100 and 300 billion cubic meters, five with reserves between 30 and 100 billion cubic meters, and 16 with between 10 and 30 billion cubic meters.[10]

In the Mendeleev Ridge area, an expedition on board the "Academic Fedorov" showed the oil reserves of this basin were estimated to have a volume of five trillion tons, which is twice that of the Shtockman deposit in the Barents Sea.

Currently, Russian explorations in the Arctic have discovered 32 large oil/gas shelf fields. Ten are already proven, including in Shtokman, Rusinov Leningrad oil fields in the West Arctic, in Pechor Sea and near Sakchalin.[11] Recently it was announced that the Russian oil company Sevsapneft was going to begin commercial exploration of the Prirazlomnoe oil field in the Barents Sea in 2005.[12]

The importance of continental shelf delimitation in the Arctic will be progressing following the exploitation level of natural resources. The first step has been made.

References

Jackson, H.R., and J. Sweeney. 1993. Alaska to Norway: Global Geoscience Transect 10, Geological Survey of Canada, Dartmouth NS. Published under the auspices of the American Geophysical Union.

Jackson, H.R., O. Eldholm, A. Grantz, W. Jokat, Y. Neprochnov, and M. Sorokin. 1996. Arctic Ocean seismic refraction catalogue. GSC Open File 4111, Geological Survey of Canada, Dartmouth NS.

Kazmin, Y., and R. Macnab. 1996. The continental shelf beyond 200 nautical miles in the Arctic Ocean: a Scientific/Technical Workshop, in Appendix C, Macnab and Nielsen, 1998.

Macnab, R., and G. Grikurov. 1997. Arctic Bathymetry Workshop, St. Petersburg, Russia, September 18-19, 1997. GSC Open File 3569, Geological Survey of Canada, Dartmouth NS.

Macnab, R., and A. Nielsen. 1998. IOC/IASC/IHO Editorial Board for the International Bathymetric Chart of the Arctic Ocean; Report of Meeting, Copenhagen, October 19-20, 1998. GSC Open File 3713, Geological Survey of Canada, Dartmouth, NS.

Sorokin, M. 1994. Geophysical investigations on the ice supported by aircraft. Proceedings of the InterRidge Workshop on Arctic ridges, GEOMAR, Kiel, Germany, pp 7-8.

Sorokin, M.Y., Y.Y. Zamansky, A.A. Kartelev, A.M. Popkpv, V.S. Golubkov, V.Y. Glebovsky, S.P. Maschenkov, and A.V. Bulatkin. 1995. Delong Islands - Makarov Basin Geotransect: preliminary mapping and interpretation results (abstract only). 21st Assembly of the International Union of Geodesy and Geophysics, Boulder CO.

Sorokin, M.Y., V.Y. Burmin, S.S. Drachev, L.R.Merklin, Y.P. Neprochnov, A.V. Elistratev, Y.G. Kiselev, H.R. Jackson, and R.F. Macnab. 1996. Recovery and digital preservation of seismic reflection and refraction data collected by Russian agencies in the Arctic. European Geophysical Society General Assembly, The Hague, The Netherlands, May 6-11, 1996. Part 1, Supplement to Volume 14, Abstract Book, p. C201.

Jones, Meirion T, 1997. The GEBCO digital atlas; a seamless, digital, bathymetric contour chart of the world's oceans. Hydro International. 1; 5, pp. 46-47.

Hare, R. and D. Monahan, 1993. A modern quantification of historic hydrographic data accuracy. *Lighthouse,* no 48, pp. 1-14.

Hughes Clarke, JE, 1999. Present day methods of depth measurement. in Cook, P, R T Haworth and Chris Carleton *eds*, Sovereign Limits beneath the Oceans: The scientific and technological aspects of the definition of the continental shelf. IOC, in press

International Hydrographic Organization, 1998. Status of hydrographic surveying and charting worldwide. Special Publication No 55, Monaco, 84pp.

International Hydrographic Organization, 1998. IHO Standards for Hydrographic Surveys, IHB, Special Publication No 44, 4th Edition.

Monahan, David and Larry Mayer, 1999. An examination of publicly available bathymetry data sets using digital mapping tools to determine their applicability to Article 76 of UNCLOS. This volume.

Monahan, David, 1999. The Interpretation of Bathymetry in Cook, P, R T Haworth and Chris Carleton *eds*, Sovereign Limits beneath the Oceans: The scientific and technological aspects of the definition of the continental shelf.

Notes

[1] Although it was decided by the eleventh Meeting of States Parties to the Convention to modify the starting date of the ten-year period required by article 4 of Annex II to the Convention for the submission of the particulars of the outer limits, the submission of the Russian Federation has been made to the Commission well within the time limit prescribed from the entry into force of the Convention for that State.

[2] The Commission was elected following the entry into force of the United Nations Convention on the Law of the Sea. Its purpose is to facilitate the implementation of its provisions in respect of the establishment of the outer limits of the continental shelf beyond 200 nautical miles from the coast. Article 76 provides the rules by which coastal States may establish those outer limits. The tasks of the Commission are twofold: to examine the submission and make recommendations to the coastal State, and to provide scientific and technical advice, if requested by the coastal State concerned during the preparation of that submission. The Commission's recommendations and actions are without prejudice to the delimitation of boundaries between States with opposite or adjacent coasts.

[3] More than 30 States are said to possibly meet the requirements to take advantage of those provisions. Commission on Limits of Continental Shelf, at headquarters, Press Release SEA/1767, 28 April - 2 May.
<http://www.un.org/Depts/los/clcs_new/commission_submissions.htm>

[4] The determination of the outer limit of the continental shelf of States is necessary to separate those areas that fall under national jurisdiction from those areas of the seabed which were proclaimed by the General Assembly, and later the United Nations Convention on the Law of the Sea, to be the common heritage of mankind. The resources of the deep seabed beyond the limits of national jurisdiction are to be managed jointly by all States through the International Seabed Authority, a body also established by the Convention.

[5] David Monahan. Achievable uncertainties in the depiction of the 2500m contour and their possible impact on continental shelf delimitation. Proceedings of the international conference on technical aspects of maritime boundary delineation and delimitation (including UNCLOS Article 76 Issues) International Hydrographic Bureau. Monaco. 8-9 September 1999, pp. 261.

[6] Institute for Geology and Mineral Resources of the Ocean (VNII Okeangeologia); Head Department of Navigation and Oceanography of the MOD; Polar Marine Geological Expedition (PMGE) and some others.

[7] Bathymetry and deep structure of the arctic continental margin of Russia in the context of article 76 of the UN Convention on the Law of the Sea. By Georgi CHERKASHOV,

I.S. GRAMBERG, A.P. MAKORTA, V.D. KAMINSKY, G.D. NARYSHKIN, V.A. POSELOV and M. SOROKIN. Proceedings of the international conference on technical aspects of maritime boundary delineation and delimitation (including UNCLOS Article 76 Issues) International Hydrographic Bureau. Monaco. 8-9 September 1999, pp. 257-258.

[8] Bernard COAKLEY. Contribution of the SCICEX project towards the implementation of article 76 of the UN Convention on the Law of the Sea in the Arctic Ocean. Proceedings of the international conference on technical aspects of maritime boundary delineation and delimitation (including UNCLOS Article 76 Issues) International Hydrographic Bureau. Monaco. 8-9 September 1999, p. 259.

[9] Statement made by the Deputy Minister for Natural Resources of the Russian Federation during presentation of the submission made by the Russian Federation to the Commission, made on March 28, 2002. United Nations Convention on the Law of the Sea. Commission on the Limits of the General Continental Shelf New York, March 25 to April 12, 2002.

[10] Date of the publication: January 18, 2001. Petroleum Potential of Shallow-Water areas of Russian Arctic Internet Geology News Letter No. 96, May 7, 2001.

[11] RIA "Novosti.". 29.04.02. http://www.opec.ru/news_doc.asp?d_no=23530.

[12] *Expert North –West*, № 336 (2004).

PANEL III: SCIENTIFIC ASPECTS OF ENERGY EXPLORATION AND EXPLOITATION AT SEA

The Concept of the International Transfer of Good Practice as an Environmental Policy Component in Major Offshore Oil and Gas Developments: A Perspective from Environmental Science

William Ritchie[*]

ABSTRACT

Offshore oil and gas developments have three main stages – exploration, production (including transport) and decommissioning; all require environmental assessment. Nation states have various requirements for the process and documentation of environmental assessment. The phrase "transfer of good practice" is often encountered, especially in newer developments in areas with relatively little prior experience with major development. This transfer can be justified on the basis of the commonly used phrase "no need to reinvent the wheel." Both concepts – "good practice" and "transfer of similar experience" – are commendable, but both need adaptation and qualification in their application to new areas, especially those with different bio-climatic and physiographic conditions. Their modification relies on two main agencies: international consultancies and in-house expertise in the multi-national oil companies. Governmental agencies in the recipient countries need to be able to match this experience and expertise to evaluate the proposal and, in time, regulate and monitor the development. It is also possible that international loan and developmental capital providers need similar assurances.

This review addresses both the transfer of generic good practice and the need to adapt planning and regulatory requirements to the reality of local and regional environmental and ecological circumstances. Nevertheless, a balance has to be achieved between unilateral local initiatives and the goal of acceptable common international standards of environmental care and protection.

[*] Adjunct Professor, World Maritime University, Sweden, and Director, Aberdeen Institute of Coastal Science and Management, Aberdeen, Scotland.

Contemporary Concerns

In the Daily Telegraph on April 17, 2004, an article appeared under the title "Oil Project is death knell for whales'" with the sub-heading "a £6 billion scheme in Russia's Far East, backed by Shell, could cause ecological damage in the Pacific." The focus was the small remaining population of the endangered Pacific Grey Whale. The project in question is the next phase in the Sakhalin oil and gas development, which has been operational since 1999. Other issues of ecological concern related to terrestrial populations of various species and threats to pre-existing ways of life. Other problems were associated with the construction of one of the largest liquified natural gas plants in the world in Aniva Bay. The possibilities for damage, including oil spills in the treacherous La Perouse Strait that lies between Hakkaido and Sakhalin, were also mentioned. The protesters from "around the world" were described as "environmental groups" and "conservationalists" (terms that may require definition), and the locus of their public protest was not the corporate headquarters of Shell in London, but the head office of the European Bank for Reconstruction and Development; a bank established with taxpayers' money to bring about the reconstruction of Russia and Eastern Europe. The developers, including Shell, Mitsubishi and Mitsui, have formed a consortium called Sakhalin Energy. This group was seeking financial assistance for this huge investment for the exploitation of a very large source of fossil fuel. Irrespective of the merits of any of the arguments, the key quotation, as reported, came from the European Bank for Reconstruction and Development: ". . . the concerns raised by the environmental groups were 'valid'. The bank was seeking more information from Shell and would be unlikely to decide whether to approve funding, and under what conditions, until the end of the year."

This clash between groups who protest strongly against oil and gas exploration and production typifies situations occurring with increasing frequency at various scales: local, national and international when any form of onshore and offshore exploitation of hydrocarbons is proposed. The Sakhalin case, however, whilst not unique, has provided a good example of the platform for protest being directed at an international funding bank in addition to the oil companies. The public headline is couched in emotive language with an initial focus on a particular species or habitat. Frequently the articulation of a protest is from a group or society usually termed "green" and sometimes "green activist." Once work commences, it is less common to see any reporting of the views of the governmental agencies and ministries that are charged, in the public interest, to care for the environment and, ultimately, to permit and to regulate the operators. When the development is offshore (and the overwhelming majority of developments are now offshore), there is added complexity in the legal and planning framework. Implicit in the reply from the bank is that any similar development anywhere in the world will be required to demonstrate adherence to international standards of good environmental practice at all stages of the project. The same source summarized this position, saying, "We are certainly not going to provide finance until we see this project complies with our environmental standards." Thus, any notion that the dialogue is only between the oil company and the competent regulator is increasingly invalid. Three or four parties are now involved.

At a general level, Sakhalin and other global examples have to satisfy multiple regulatory and compliance criteria at regional, national and international levels (and engage in massive and costly public relations) in order to counter the threat, both real and perceived, of damage to the environment and of serious disruptions to pre-existing economies and ways of life, especially if these are traditional or dependent on natural resources. Although many other issues need to be addressed, including various codes of law, such as maritime law, and ethical concepts, such as sustainability, and, on the other hand, the undeniable global need for energy from oil and natural gas, increasingly, wide-ranging environmental conditions and requirements will be placed upon the developers (and their funding partners). The key players have become more dependent on a thorough understanding of the nature and reliability of environmental science, including the easy to say but difficult to define concept of international good practice.

THE REGULATORY DOMAIN

For a major international development, especially one that is coastal or marine, the list of statutes, laws, protocols and conventions seems endless. It is also increasingly layered. An example would be a U.K. offshore development in the Atlantic Ocean, which would need to consider international maritime law; international conservational protocols; European laws and conventions; U.K. planning and other legal requirements, such as those relating to pollution; and governmental regulatory departmental rules, including those made by the Department of Trade and Industry. Some developments might include transboundary concerns. Since almost all such developments involve transport both by vessels and by pipelines, complex maritime and shipping laws will also apply. Moreover, whether or not the oil or natural gas comes ashore, the risk of an oil spill reaching some coastline will need to be assessed by local authorities and agencies that must now be added to the list of interested parties who can legally influence the decision-making process. Within this long list, and depending on time and place, an offshore development exemplifies the concept of relativities and possible inconsistencies of treatment in that different facets might receive more or less attention. The exceptional attention paid to marine pollution has been summarised by Gold (1998): "In an increasingly environmentally cautious world, ship-source marine pollution has, for a long time, been singled-out for special attention." In reality only about 10 percent of marine pollution, mainly oil, is from ships, and it is likely to be decreasing. About 50 percent of such pollution is land-based. Hence the attention is not commensurate with the risk. As quoted by Gold (1998), more than a decade ago, the Secretary-General of UNCED (United Nations Conference on Environment and Development) could say that on the basis of MARPOL, "[a]ccidental spills are relatively isolated geographically . . . transient effects in the vicinity of the accident . . . more chronic conditions in localised sites . . . petroleum pollution does not now represent a severe threat to marine habitats and organisms. However, since accidental spills cannot be totally avoided, contingency planning and effective response action are essential." For many parts of the world the position has probably improved over the ensuing decade in relation to prevention and to adherence to regulatory requirements, both in port and harbours and in sea-going vessels. The record for production platforms and pipelines would suggest similar improvement, and these are almost negligible sources of significant marine pollution (Intertanko, 2004). Commonly, as summarised by Mukherjee (2003), offshore platforms and facilities also have a less-than-clear position in international marine law.

If it is correct that ships, platforms and pipelines are now less likely to produce significant environmental damage, it seems paradoxical that legal requirements, public protests, regulatory constraints and the burden of preventative and mitigation measures are all increasing. There is also mounting evidence that, in time, major oil spills and chronic small discharges of hydrocarbons into the marine environment do little lasting damage (Sell, 1995; Abdulaziz et al., 1997; Ritchie, 2003) and, according to Holden (1990), "There is a good deal of overreaction in connection with oilspills and that the money spent on cleaning-up after could have been made better use of elsewhere."

The answer to this paradox could take two forms. First that prevention, mitigation, regulatory conditions and increasing environmental awareness at managerial levels have brought about these dramatic improvements. Thus more regulations could reduce further the risk of any damaging pollution from an offshore activity. This answer does little, however, to answer the question of whether there is a need for more stringency when there is little hard evidence that significant damage to habitats and species is occurring due to offshore oil and gas developments at present, other than the possibility of damage at a few localised spots that are extremely small in relation to the vast areas of seabed and overlying water masses. The second response to the paradox relates more to the rapid increase in the global importance of a wide range of environmental issues that are now near the top of the agenda at political and governmental levels and are equally high on the list of public concerns. Arguably the particular concerns of the oil and gas industry have been wrapped up in the growing awareness of possible climatic change and the various scenarios of significant deterioration in the health of the planet. To this extent, both politicians and the public seem to have become more intolerant of any level of pollution or risk to the environment, regardless of whether the statistical probability of an accident is very low. There is probably even less acceptance of the view that if a pollution event did occur, local ecosystems would be sufficiently resilient to absorb the change or that any changes that might occur would be less than natural variations in the condition of the habitat or fluctuations in biotic populations.

The Sakhalin example, in microcosm, contains all of the typical elements of the collision of multiple factors and strongly held attitudes, including public perception, trust in science, risk assessment, value judgments, economic benefit and political positions – and all within a regulatory domain in transition.

ENVIRONMENTAL ASSESSMENT

Irrespective of the scale, location, product and time of development, some form of environmental impact assessment lies at the core of all permitting and regulatory processes – both as a legal requirement and as a critical element in public relations. Although numerous textbooks (e.g., Glasson et al., 1998), academic articles and dedicated journals exist, it is difficult to provide a date for the emergence of this process, although the U.S. National Environmental Policy Act in 1969 is often cited as the key development – and it is necessary to emphasize that an EIA document is essentially the product of a multi-disciplinary process (O'Riordan, 2000). Some writers have noted that some concern for the environment has been embodied for centuries in Acts relating to public developments that were in the national interest, such as in the U.K. Provision of Water Supply Act, 1856, and EIA is therefore no more than a natural evolution of the importance of care, not just for public health but also for natural conditions. In the 1970s references were made to the relationship between cost-benefit and environmental assessment – a process that is probably impossible in strict economic terms because so much within environmental assessment is qualitative (e.g., cultural, ecological, historical and other value judgments). Rightly, there is a correlation between the growing environmental movement at national and international levels and specific key conference resolutions, landmark provisions such as Club of Rome (1972 and 1989); the UN Conference on Environment and Development (Rio, 1992); UN Conference on Human Environment (Stockholm, 1972); and the Brundtlant Report (1987). Much of the Kyoto (1997) protocols are of particular current relevance as a result of growing concern with the possibility of accelerated climatic change due to greenhouse gas effects. At regional rather than international levels, important national legislation has also provided milestones on the path to codifying public concern with the state of natural environments. Such legislation includes the U.S. National Parks Act and the U.K. Countryside and Wildlife Act.

Counter opinions, while not denying the need to protect the environment and to promote ecological principles as the means of achieving reasonable sustainability, have been less common. Increasingly, the term "precautionary principle" has been quoted alongside "sustainability" in legal, planning and public use whereas the long-established scientific concepts of uncertainty or statistical risk would be much more appropriate. Unfortunately, the deceptive simplicity of the short terminology, i.e., "precautionary principle," seems to have overtaken the more complex but better scientific approach of some type of probability calculation (e.g., Lomborg, 1998; Holden, 1990; Hohn and Harris, 1998; Appell, 2001; and Graham, 2004). Natural resilience and, especially, natural fluctuations in habitat conditions and population dynamics on variable timescales are well-documented and of extreme importance, but science has failed to communicate these long-standing and proven principles with both the government and the public. It is also arguable that even the most thorough and well-documented environmental impact statement rarely includes adequate presentation of trends, natural variability, confidence levels, resilience and alternative natural causal factors. (In a different sense, few environmental statements and assessments look closely at the impact of *existing* land and sea uses as factors producing negative changes to the environment; the existing uses are listed and analyzed but normally in the context of how they might be impacted.) The current impact of fishing on habitants and marine ecology in the same area, for example, is rarely discussed. This is analogous to the position described previously by Gold (1998), where shipping received considerable attention for its 10 percent contribution to marine pollution, with much less attention being given to what was causing the remaining 90 percent.

By custom and practice, typical major environmental impact assessments operate on a relatively standard matrix-type format with a check list that is often divided into two main sections: Site Specific and Activity Specific.

Typical General Format of an EIA for a Major Offshore Oil or Gas Development

Context	Including a description of the project, the benefits, the legislative framework, consultation process and areas of uncertainty.
Option Appraisal	Including a consideration of alternative sites.
The Planning Framework(s)	At various levels.
Project Description	Normally divided into different sub-projects and time phases; usually includes site preparation, construction operation and decommissioning. For a major development this can be a complex section in the EIA.
Description of Existing Environmental and Ecological Conditions	This is normally divided into a series of categories under broad initial divisions such as atmospheric, marine, and seabed, within which topics such as sediments, noise, water movement and quality, winds, seabirds, sea mammals, etc. are described. This section, even for an offshore development, might include archaeology and existing usage such as fishing. It is also usual to describe the nearest coastal

	environment on the basis of the impact trajectory of an oil spill.
Assessment of Environmental Impact	This section considers how the project will impact differentially on the checklist of the environmental and ecological attributes of the area. Most reports will also contain a further subdivision into normal and abnormal operating conditions. Emission and discharges are important elements in this assessment process. Increasingly, noise, visual effects, radiation and materials used are included in this review. A clear distinction is made between factors that are governed by external legal regulations and those that are controlled by other types of requirements. More recent assessments have tended to include reference to stakeholder and public concern. In the United Kingdom there is a requirement to consider how much additional energy is used in carrying out this work. Ultimately value judgments are made. This is the area of assessment. The judgments are classified as not significant, low, medium or high, and on this basis managerial, mitigation and preventative solutions are requested. All these headings and elements are invariably presented in a matrix format with numerical scaling. Most reports make excellent use of color coding to highlight areas of greater concern.
Highest Potential Risk of Damage	Normally the highest risk aspects from the assessment are given special attention in a separate "chapter." Cumulative effects and mitigation measures are discussed. This section is dependent on the quality and quantity of pre-existing data and information. If gaps are identified they are highlighted for action. Whenever possible, this section becomes more quantitative.
General Conclusion and Managerial Commitments	Most major oil and gas companies have well-established protocols and statements of commitment to good practice. There is therefore a clear requirement for the Environment Statement to reflect the internal controls that are and will be applied to the project. Phrases such as "best practice" or "only practicable system" will occur in this section.

Over the last two decades the different phases of development have been considered in a more integrated way, from exploration to operation to close-down. In the end, however, this matrix requires the consultant, the developer and the permitting and regulatory authorities to exert judgments that are at least partly non-quantitative. This is typified by terminology throughout the assessment process that consists of language such as light, moderate, severe and significant. In contrast, licensing authorities have become increasingly quantitative (e.g., decibel levels for seismic work, concentration levels for chemicals as discharged in production water, chemical content of gaseous emissions in flares, etc.). Environmental impact assessments must acknowledge these compliance standards and use them as feedback loops within the process of assessment. One result of this feedback between compliance requirements and the assessment process is that it ensures that technical, engineering and operational designs and plans, *a priori*, ensure that these standards can be met. The scientific reasoning and the process of deciding the acceptability of specific threshold values should be a matter of more debate, asking questions such as, "Why is the allowable discharge of hydrocarbons in water as set by a regulatory agency 15 ppm and not 20 ppm?" Yet the increasing need to quantify some operational impacts on the environment is a significant change in the methodology of conducting environmental impact assessments. Environmental Impact Statements never question the scientific basis for deciding these thresholds, but accept them as inflexible standards. From the perspective of environmental science, however, the actual impact of a specific level of discharge, for example, could vary according to the location and ambient environmental conditions, such as dilution effects due to water movements, but this only seems to be considered in a negative sense if, for example, the body of water is enclosed or low energy.

Environmental impact assessment remains the core activity in seeking permission to develop a major oil and/or natural gas reservoir in Europe and in the United Kingdom. In particular, the assessment process has moved from what could be termed reactive to partly proactive, which, in theory, reduces repetitive research and documentation. This helps the developer to identify in advance which issues are likely to be the key issues by creating Strategic Environmental Assessment (SEA) documents for the offshore licensing of natural gas and oil lease blocks.

The background of the SEA process consists of an evolution from the practice of environmental assessment as it has developed in the last decade, mainly driven by Directive 85/337/EEC, to a more legislative approach (Amendment Directive 97/11/EC). In practice, it represents a more uniform application of the broader aims of the European Union for the protection of the marine environment, which includes a shift from the traditional form of project environmental impact assessment to a more strategic level. Directive 2001/42/EC entered into force in 2001, but the United Kingdom as a member State decided to transpose this into its legislation in July 2004 (DTI). The stated objective is "to provide for a high level of protection of the environment and to contribute to an integration of environmental considerations into the preparation and adoption of plans and programmes with a view to promoting sustainable development, by ensuring that, in accordance with this Directive, an environmental assessment is carried out of certain plans and programmes which are likely to have significant effects on the environment."

In 1999 the Department of Trade and Industry began a sequence of sectoral SEAs considering the implications of further licensing of the block for oil and gas exploration and production in the U.K. Continental Shelf. According to the excellent and informative government Web site <http://www.offshore-sea.org.uk/sea>, SEAs were in line with the United Kingdom's "greening government." In effect the first SEA was conducted in 1999-2000 for the consideration of the licensing of deep water blocks between Shetland and the Faroes. Subsequent SEA documents have been completed sequentially and have reached SEA 5. The area east of mainland Scotland toward the meridian line in the North Sea will be submitted in 2004. (For a full account of the background, the process, the regulatory context and extensive related information please consult the Web site to which earlier reference is provided.) SEA 4, published in 2003, can be used as an example for its list of extensive contents (Appendix 1). The significance of the content of Appendix 1 is its comprehensive cover of almost every issue from geology to taxation. The process is also significant; each SEA is subject to review by stakeholder groups and can be amended and improved by these consultations so that the final version is as up-to-date and as accurate as possible. More important, however, is the underlying fact that although triggered by a specific type of development, such as oil and gas licence blocks, SEA documents are proactive, not reactive, and have the potential to address other types of marine exploitation. They set the scene and help potential developers understand the sensitivities and environmental priorities. As a comprehensive survey, SEA documents also have the value of avoiding duplication by major oil companies that must invest their own resources in obtaining a similar store of environmental and related information, thereby helping concentrate research by their own staff and consultants on priority areas. Since the SEAs are wholly funded by government they are entirely in the public domain. What remains to be tested is their effective legal status. Having been endorsed by their publication after review by the government and its agencies, are they sufficiently definitive to be used by a developer to counter an appeal from an individual or body who questions the proposal under a general application of the precautionary principle or for some other fear or specific concern, for example for a habitat or species? Although explicitly linked to the regulatory process by the Department of Trade and Industry (DTI) as Licensing Authority for offshore exploratory and production leases, the objectives as described use terms like "to contribute to" and "with a view to promoting"; phrases that imply to someone who is not a legal expert an advisory or consultative status. Nevertheless, its existence in advance of a proposed development would appear to confer both considerable status and force that might, to some extent, preempt objections on ecological and environmental grounds. Perhaps, as in other matters in law, case law needs to develop before an opinion can be safely expressed.

SEAs do not undermine or replace project-specific environmental impact assessments, which will remain mandatory under both European and national legislation and regulatory requirements. Thus, SEAs have re-emphasized the importance of a sound science and knowledge base, both in the positive sense of providing a reasonable review of available information and data and, by default, exposing areas of possible environmental concern where more work needs to be done.

Some Aspects of the Nature of Relevant Environmental Science

The definition of relevant environmental science is a subject of vast debate and discussion. As analyzed comprehensively by O'Riordan (2000) there are many interpretations of the value of science, especially environmental science in its application to environmental and ecological problems. Communication of science and the politics surrounding its use are equally complex and variable. What is clear, however, is the fact that an environmental impact assessment is now much more than earth sciences and life sciences. It increasingly involves social sciences and some humanities such as ethics, although as quoted by O'Riordan (2000), the Royal Commission on Environmental Pollution (1998) stated, "A clear dividing line should be drawn between analysis of scientific evidence and consideration of ethical and social issues, which are outside the scope of scientific assessment."

All of these complex issues and their underlying concepts, philosophies, definitions and meanings are rarely part of the more pragmatic process of determining if a major offshore oil and gas development is given approval, albeit with restrictions and conditions, by the regulating and permitting authorities of a nation or coastal State. Nevertheless, the international processes that are now applicable integrate and subsume the multiple conceptual and philosophical strands, legal pressures and social attitudes that have fused together over a long period of time. When the first offshore oil was produced in the Gulf of Mexico in 1938, the public interest was probably confined to technical innovation and economic benefit with little thought to environmental consequences. In 1938, in contrast to the current plethora of regulations and requirements, including post-production decommissioning and clean-up, early offshore oil production was an exciting part of the drive to exploit natural resources for profit or employment against a background of accelerating consumer demand. Thus, although the argument might be extreme, the regulatory burdens on the exploitation of oil and gas resources now reflect the contemporary summation of knowledge and ignorance, prejudice and concern, good science and irrational fear that seem far-removed from the ethos and criteria of the 1930s. Curiously, somewhere in this process of change, any recognition of the advances in technology in exploiting and transporting the resources of natural oil and gas seems to have diminished. Large branches of technological and operational expertise have been diverted from production to the prevention of damage, to clean-up techniques and to mitigation of effects, all of which ultimately register as a cost for the project. In the example of the Sakhalin "problem," there is little mention of whether or not the oil and gas that will be extracted from the seabed can be transported by pipeline and ships, and can be stored, processed and re-exported; the complex technology that will be needed to complete the project is not questioned. The issue of prime concern is the risk of environmental damage. Recent developments in deepwater in the Atlantic Frontier (AFEN, 1998) suggest a paradox in public perception. The incredible ability of engineers to drill and extract oil from complex reservoirs with sophisticated production facilities on the deep sea floor is almost taken for granted, but the same trust is not placed on the same companies to be able to provide technological and managerial solutions to an oil spill accident or to prevent discharges at levels that allegedly exceed the carrying capacity of the adjacent seabed and water column. Does this reflect the traditional public attitude that perceives a strong predictability and confidence in hard science and technology but finds it difficult to grasp the relatively higher intrinsic uncertainties of ecology and earth sciences? Or is it a manifestation of what O'Riordan (2000) describes as the "power shift", where the "burden of proof has moved from the victim to the developer"? Environmental risks that were acceptable in the past are now unacceptable or, in terms of probability, the 1:1,000 chance is no longer acceptable and needs to be increased to 1:10,000 or more – and in some instances, to the impossible target of zero risk – with the burden of proof on the developer. This extreme interpretation has not as yet been reached. According to Hughes et al. (2002), the precautionary principle that could be labeled "the principle of uncertainty" has evolved in United Kingdom though three stages:

1. cautious progress until a process/project is judged "innocent";
2. ordinary progress until findings of guilt are made; and
3. no progress until intensive research has been conducted into a proposed process and its innocence has been demonstrated, this being the strongest interpretation of the precautionary principle.

In the subsequent exposition of the interpretation of the "precautionary principle," Hughes et al. (p. 25) quotes three prime sources: "Sustainable Development in the UK Strategy"; the White Paper "The Common Inheritance"; and the Rio Declaration of 1992, which includes reference to "scientific knowledge," "judgement," "serious threats," "lack of full scientific certainty" and, most important, "remains committed to basing action on fact, not fantasy, using the best scientific information available: precipitate action on the basis of inadequate evidence is the wrong response." Again, as part of the argument that there is a strong level of subjectivity in the process of environmental prediction of the consequences of a proposed development, the following quotation is highly significant: "[L]ikewise predictions as to the consequences of a particular process may have a strong element of subjectivity in them.

There may be initial disagreements as to the test parameters to be used in the initial collection of data and once it has been collected it needs evaluation, and hence a further departure from strict objectivity." Thus, the scene is set for the definition of the arbiters of quality and standards. Who assesses the significance of the environmental evidence upon which regulatory decisions, even go or no-go decisions, will be made? How is international good practice delivered?

PROVISION OF INFORMATION

If international comparisons are required, the first stage of inquiry must include an audit of the nature, source and validity of the information upon which decisions are made. For most environmental impact assessments the mix of data sources and their quality are extremely variable. The best source is pre-existing data and surveys that were not obtained for the purpose of considering the impact of offshore oil and gas developments. If the statistics are a long data set, it is even better. Some areas of the world have relatively good data on bathymetry, currents, seabed sediments, wave and tidal conditions. Some rely on internationally accepted services, such as the U.K. Hydrographic Office. Weather statistics are also available at a regional level. Biological and ecological information is extremely patchy and ranges from zero to sample, spot or transect surveys at best. The relative lack of oceanographic data is well known, and although improving rapidly, remains relatively weak and, moreover, is very expensive to acquire. Satellite information is helpful, especially for sea surface conditions and sometimes plankton distribution, but it has limited value in studies relating to the development of an area as small as a licence block or similar area. Digging down into raw data and information is very expensive and time consuming. In general, therefore, there is a tendency to accept all published information as being of good quality and reliability. From this point of having uncritical acceptance of outcomes, the chances of extrapolation and over-generalization become real. The original caveats and health warnings tend to be omitted. Statistical variability and error margins might also disappear. The importance of scale and grid size is rarely discussed – for example, maps of distribution might be based on original surveys within 10 km square units or, worse, by random sampling, but subsequent maps might show continuous distributions with apparently convincing boundaries. Many maps are based on isolines, which are constructed from interpolations between very few data points. Nevertheless, these pre-existing sources of information, whether scientifically strong or weak, and irrespective of the provider – the developer, the consultant, the government agencies – are the best that are available and must be used. There should be no doubt that the quality of these end sources is variable in different world locations. Put another way, for some parts of the world spatial information is at the scale and reliability of an atlas, whereas for others, maps and charts might be at a useful scale of 1:50,000. Unfortunately, when these large and small scale maps are converted to GIS displays, the visual impression is stunningly attractive and apparently definitive, but questions as to where, when and how the spatial data was obtained for the GIS compilation are rarely asked.

At a more local level, information is normally serendipitous. Good fortune might provide a series of useful papers in academic journals. Local research institutes (both national and university based) might have done detailed surveys of habitats and species. Oceanographic vessels might have traversed the area in question. If a nation state has anticipated possible offshore developments, direction might have been given to undertake appropriate research in potential areas. This process has been described by the general term "foresight investigation." Sometimes arguments based on "analogous areas" might be used. This can be described best by a hypothetical example: If the EIA reveals that a pipeline will cross a seabed area with ecologically important seagrass beds and no survey work has been done in this specific area, the literature will be scoured to find a scientist who has worked on similar sea grasses in comparable areas to express an opinion with or without a site visit. Most experts will insist on a site visit and a reconnaissance survey at the very least, but time and cost might be prohibitive. This technique is almost universal and the expert will usually be introduced into the program by the developer or the consultancy company undertaking the EIA after the first environmental scoping exercise, which usually precedes the full EIA investigation. This is a reasonable procedure used in almost all areas of science within the EIA. The experts who give the advice are well aware of the imperfections and risks attached to this methodology but accept that this is likely to be the best evidence that can be provided.

Inevitably, a limited number of impacts will be classified as significant and will demand additional information. Supplementary surveys and measurements might be necessary. These special surveys take time. Two types of survey are common. Some are part of the inventory of static conditions, such as geology, and some require information on dynamic changes. It is normally impossible for processes to be measured on anything other than a short-term basis. Rates of change, variation, sampling density, independent controls, and space and time constraints all reduce the scientific value of the measurements. Most surveys are therefore indicative rather than definitive. If there are seasonal variations, as with most biological populations, great care needs to be taken. It has been shown, for example, in a biological survey near Singapore (Lee and Sin, 2003) that if the same survey had been replicated two months later the results of a botanical survey would have been significantly different. In some surveys, such as beach and dune developmental processes, static information such as coastal landforms (Ritchie and Gimingham, 1989) is used to deduce wave- or tide-generated dynamic factors. In almost all instances, the criticism of "quick" surveys is the inadequacy of sampling and the degree of representativeness. Almost all such surveys contain warnings and statements that more time is needed to assess the validity of the survey – but time is rarely, if ever, available and the warnings have to be overruled. Hence this information falls into the category of "best that is available." These survey inadequacies are not solely within the domain of the developer, but are equally, if not more frequently, part of typical arguments and reasons given by objectors who oppose the development. It is also important to note that this type of environmental data – pre-existing or commissioned – will form the database for subsequent monitoring and rehabilitation and restoration at the end of the life of the project.

Setting the International Standard

From a scientific perspective the typical requirement of an EIA for offshore oil and gas developments could be constructed in several ways, including the following:

(i) static and dynamic information;
(ii) biotic and abiotic information;
(iii) information deficits at regional and local scale; and
(iv) direct and indirect impact.

Some statements of trends and variability, including the statistically defined degree of risk and uncertainty, should be included. Some factors are uncontrollable and will occur whether or not any development of any kind takes place (e.g., weather, ocean currents, tidal patterns); such factors should be defined as contextual, and although extremely important, are beyond the power of any developer to alter or modify. In reality the common type of assessment, described previously as the generic Environmental Impact Statement, does not use this format. The traditional approach remains as a checklist of standard environmental factors and conditions rather than a more process-based system, but for the most part, this is a taxonomic problem rather than deliberate obfuscation and reflects the history of their production. In effect, the content and presentation have become formalized and standardized, which is probably advantageous when international comparisons have to be made. Different formats and national requirements could produce considerable difficulties for third-party international funding agencies or overseeing regulatory bodies when they are required to assess the quality and relevance of EIAs. On the assumption that the "contents list" is unlikely to change, the key question remaining is who provides the scientific evidence, who judges its importance, and who provides independent checks. The separate question of who or what defines and justifies standards and critical values is an equally important issue and requires much more research and reflection.

If international comparisons have to be made, the available options, either individually or collectively, are as follows:

1. A single independent expert of international standing and experience with no previous connection with the project;
2. A panel of independent experts as defined above;
3. An internationally recognised and acceptable environmental consultancy company;
4. A panel of experts set up by the regulating agency and/or government; and
5. A panel of experts set up by the developer(s).

Little experience is available to provide any advice on how the above options can occur. Within the components of an EIA or an application to develop offshore oil and gas assets, individual expertise can be obtained to check and to validate one or several aspects of the environmental and other types of assessment data and interpretation. Such expertise cannot normally interrogate raw data or original pre-existing research, but it can give a reasonable opinion of the scientific processes and procedures that have been followed. Reports can be read and scientific references cross-checked. In another sense, the procedures that have also been followed, meaning the sequence of developing the arguments, also need to be verified. The difficulties that can arise relate more to who appoints and who pays for these independent checks. The affiliations of the expert could also come under scrutiny. The same arguments apply to a panel or group of experts. A reasonable proposition is for all the main parties to agree to fund this review jointly. Another suggestion would be to use university or research centers or national academies or institutes as contractors. There is no reason why this could not be an international effort. The crux issues are credibility, confidence and trust. Technical and scientific expertise is also vital, and clear terms of reference that do not include any possibility of negotiation must be provided from the outset.

This process of independent expert review, however, might engender hostilities and provoke understandable sensitivities on such matters as the competencies of governmental bodies and agencies to make their own decisions using their own expert advice and opinion. The developer might also have reservations about the implied lack of acceptance of its own EIA assessments, whether produced in-house or by consultancy companies. Both parties might be skeptical of the opinion of a science-based expert group with little knowledge and experience of commercial and operational realities. The time schedule for this individual or group to produce a report would be problematical: If the amount of time was too short, no one would believe it had done the work thoroughly; too long and the costs associated with delays (sometimes referred to as planning blight) could escalate. From a detached perspective, it seems unfortunate that an impasse might be created to the extent that independent arbitration becomes the only rational solution. Nevertheless, given the increasingly bitter collision between almost every aspect of oil and gas exploration and production worldwide and opposing interests ranging from small public protest groups, often with very narrow concerns, to national and international regulations and guidelines that are increasingly demanding, the future looks less than harmonious. For major offshore and coastal oil and gas developments, the Sakhalin situation is likely to be the norm and not the exception.

Appendix 1
CONTENTS LIST

1.6 Organisation Of The Consultation Document

The consultation document comprises 12 Sections with a glossary and a non-technical summary. Figures and tables are interspersed throughout the document.

The **non-technical summary** is intended as a comprehensive standalone summary of the SEA, its findings and conclusions.

Section 1 Introduction and Background provides both a context and guide to the main body of the report.

Section 2 Strategic Environmental Assessment Process provides an overview of the various stages and activities leading up to this public consultation phase.

Section 3 Regulatory Context summarises the requirements of the SEA Directive, the oil and gas licensing process together with an overview of environmental legislation and controls in relation to the oil and gas industry offshore.

Section 4 Activities describes the alternatives to the proposed action and the activities arising (and more fully described in a supporting document, SD_002, available on the SEA Web site).

Section 5 Physical and Chemical Environment describes the geology, sediments, climatic conditions and oceanography of the area, together with a consideration of the existing levels their sources.

Section 6 Ecology addresses the biological features of the area together with their ecological importance and sensitivity to oil and gas activity.

Section 7 Coastal and Offshore Conservation specifically considers habitats and species of relevance in the context of the *Offshore Petroleum Activities* (*Conservation of Habitats*) *Regulations, 2001.*

Section 8 Users of the Sea and Coastal Environment describes the commercial and other human interests and activities in the coastal and offshore area.

Section 9 Other European Resources of Potential Relevance to SEA 4 summarises coastal resources and conservation interests in these areas.

Section 10 Consideration of the Effects of Licensing describes the method used to screen potential effects together with a more detailed consideration of those environmental interactions with the potential to cause significant effects and including cumulative, synergistic and transboundary effects. Mitigation measures are also considered.

Section 11 Conclusions provides an overall conclusion regarding the likely implications of the proposed licensing and alternatives, together with recommendations for mitigation and monitoring and gaps in understanding relevant to the process.

Section 12 References lists the data sources used in the conduct of the SEA4 and referenced in the Report.

Note: CD versions of SEA Reports can be obtained by contacting the consultancy contractor to the Department of Trade and Industry in the U.K. (DTI) at the following address:
<http://www.hartleyanderson.com>

Elements Described in Section 5 Above
Section 5

1. Physical and Chemical Environment
2. Geology Substrates and Shoreline Types
3. Climate and Meteorology
4. Hydrography
5. Contamination
6. Ecology
   - Plankton
   - Bentho

- Cephelapods
- Fish and Shellfish
- Birds
- Marine Mammals

BIBLIOGRAPHY

Abdulaziz, A and Krupp, F	1997	What Happened to the Gulf: Two years After the World's Greatest Oil Slick. Arabian Wildlife 2, 1
Appell, D	2001	The New Uncertainty Principle. Scientific American. January Edition.
Brundtlant Report	1987	Our Common Future World Commission on Environment and Development
Club of Rome	1972	The Limits to Growth Meadows. D.L. et al. University Books. New York
Club of Rome	1989	Beyond the Limits to Growth. Pestel. E. University Books. New York
Gold, E	1998	Gard handbook on Marine Pollution, 2nd Edition. Foreningen Gard. Arendal. Norway 460pp.
Glasson, J, Thevivel, R and Chadwick, A	1998	Introduction to Environmental Impact Assessment U.C.L. Press, London
Graham, JD	2004	The Perils of the Precautionary Principle: Lessons from the American and European Experience. Heritage Lecture #818, Jan 15, 2004. USA
Hohn, S, and Harris, J	1999	Precautionary Principle Stifles Discovery. Nature 400 : 398
Holden, C	1990	Spilled oil looks worse on TV Science, 250, 371
Hughes, D, Jewell, T, Lowther, J, Parpworth, N, de Prez, P	2002	Environmental Law (4th Ed) Butterworths, London, 725pp
Intertanko	2004	Tanker factsheet, London 2004
Kyoto	1997	The Kyoto Protocol United Nations. Kyoto

Lee, M and Sin TS	2003	Marine Environmental Baselines: Design considerations for monitoring and impact assessment. Proc. Int. Conf. On Port and Maritime R & D and Technology. Singapore. Vol.2. 31-37
Lomborg, G	1998	The Skeptical Environmentalist Cambridge University Press
Mukherjee, PK	2002	Maritime Legislation. World Maritime University. 353pp
O'Riordan, T	2000	Environmental Science for Environmental Management, 2nd Edition. Prentice Hall, London, 520pp.
Rio	1992	United Nations Framework Convention on Climatic Change
Ritchie, W	2003	Chronic, Small and Major Oilspills and Discharges – Occurrence and Environmental Costs: The Case Study of Sullom Voe, Shetland Islands. Proc. Int. Conf. Ports and Maritime R & D and Technology 2, 7-18, Singapore.
Ritchie, W and Gimingham, CH	1989	Restoration of Coastal Dunes Breached by Pipeline Landfalls in North-East Scotland. Proc. Roy. Soc. Edinb. 96B, 231-245
Royal Commission on Environmental Pollution	1998	21st Report. Setting Environmental Standards. Cm. 4053. H.M.S.O. London.
Sell, D, Baker, JM, Dunnet GM, McIntyre, AD, and Clark, RB	1995	Scientific Criteria to Optimise Oil Spill Clean-Up. Proc. Int. Oil Spill Conference, Long Beach, California 595-610
U.S. Government	1969	National Environmental Policy Act

Challenges of Collecting Law of the Sea Data in the Arctic

Larry Mayer,[*] *Martin Jakobsson*[†] *and John Hall*[‡]

INTRODUCTION

Article 76 of the United Nations Convention on the Law of the Sea (UNCLOS) offers the opportunity for coastal States to claim, within limits, jurisdiction over the seabed and subsoil of "submerged extensions of the continental margin" beyond their 200 nautical mile Exclusive Economic Zone (EEZ). To take advantage of this opportunity, however, a coastal State must submit to the United Nations Commission on the Limits of the Continental Shelf (CLCS) detailed documentation that demonstrates certain criteria have been met. These criteria, as described in UNCLOS Article 76, involve the geodetic calculation of distances from territorial baselines, measurements of the depth and the shape of the seafloor, as well as determinations of the thickness of sediment accumulated on the margin. As such, a claim made under UNCLOS Article 76 involves the collection and interpretation of geological and geophysical data, not unlike what is done in support of the exploration for offshore resources and, in particular, the exploration for hydrocarbons. Given the generally broad, resource-rich continental shelves associated with the Arctic Ocean, there is great potential for claims in the Arctic under UNCLOS Article 76 for at least five nations: Canada, Denmark, Norway, Russia, and the United States. But given the harsh environment and often ice-covered waters of the Arctic, the challenges of collecting the required data are also great.

DATA REQUIRED

Under UNCLOS Article 76, a coastal State may claim jurisdiction over the seabed and subsurface of submerged extensions of their continental margins if it can be demonstrated that the margin is a "natural prolongation" of the state's territorial landmass. While this demonstration (deemed the "test of appurtenance" by the CLCS) can, in many cases, be

[*] Director, Center for Coastal and Ocean Mapping, University of New Hampshire.
[†] Center for Coastal and Ocean Mapping, University of New Hampshire.
[‡] Center for Coastal and Ocean Mapping, University of New Hampshire.

quite straightforward (e.g., when a coastal State has a broad, shallow continental shelf and/or a very thick adjacent sediment column), in other situations this demonstration is subject to a detailed understanding of the geological and geophysical origins of the margin. We will not deal with these interpretive issues in this paper, but rather focus on the data required for making a claim under Article 76 once the test of appurtenance has been passed.

Having demonstrated that their margin is a natural prolongation of the territorial landmass, a coastal State then must collect (or use previously collected) data to establish a series of "formula" and "cutoff" lines, the combination of which determines whether a claim can be made beyond the 200 nm EEZ and, if so, the extent of that claim. As outlined under UNCLOS Article 76, a claim can be extended to a line connecting points that are 60 nm beyond the foot of the continental slope and/or a line connecting points where the sediment thickness under the margin is 1 percent of the distance back to the foot of the slope (the Gardiner Line). These are the formula lines, and they can be combined in a way that is most advantageous to the coastal State. The extent of the claim is limited however, by the cutoff lines, which are either lines 350 nautical miles from the territorial baseline or lines 100 nautical miles seaward of the 2,500 m depth contour. Again, these cutoff lines can be combined in a way that is most advantageous to the coastal State. A coastal State also has the option of using "evidence to the contrary" – bringing in other geological or geophysical evidence that might support a claim of a natural prolongation of the territorial landmass that differs from that defined by the simple formulae lines.

Thus, in order to establish a claim under UNCLOS Article 76, a coastal State must make measurements of the absolute depth of the seafloor (to determine the 2,500 m contour), the shape of the seafloor (for determination of the "foot of the slope"), the distance from the territorial baseline, and the thickness of the sediment column. While the distance measurements are simple calculations, once official territorial baselines are established, measurement of the absolute depth and shape of the seafloor (bathymetric measurements) and measurement of the sediment thickness (from seismic profiling or drilling) involve often logistically complex field programs. Other geological and geophysical evidence, such as gravity and magnetics data, geologic mapping, boreholes, and well-logging may also be useful, particularly if a coastal State is calling upon "evidence to the contrary." While techniques are well-established for the collection of all of these types of data, their collection in the Arctic presents special challenges.

How are these data collected?
Bathymetry

Information on the depth of the navigable waters has been collected for thousands of years as ancient mariners would toss a weighted line ahead of their craft to prevent running aground. The "lead line" remained the only mechanism to measure depths until well into the 20th century. In deep water, lead lines are very time consuming and quite inaccurate. Thus, until the 20th century our understanding of ocean depths was very limited. After the Second World War, the echo sounder, mounted on a vessel, became commonly available and provided a relatively rapid means of measuring depth, even in deep water. Until about 25 years ago echo sounders used a single broad beam of sound to make a single measurement of depth at any one spot. The sound spread laterally as it approached the seafloor and thus the single depth measurement was an approximation of the depth over a large lateral area.

Over the past 25 years, multi-beam sonars have been developed, revolutionizing our ability to image and map the seafloor. A multi-beam sonar produces a broad swath of beams that make many (often hundreds) of simultaneous measurements of depth across a wide area of the seafloor. As opposed to the relatively inaccurate, sparse measurements provided by lead lines, or the two-dimensional profiles provided by single-beam sonars, the multi-beam sonar produces a complete and accurate three-dimensional picture of the seafloor morphology (see Appendix, Figure 1, a comparison showing leadline, single-beam and multi-beam). While multi-beam sonar data is not required in order to make a claim under UNCLOS Article 76, it has been demonstrated (Mayer, et al., 2002), that the use of multi-beam sonar data in combination with modern, interactive three-dimensional visualization techniques can remove much of the ambiguity associated with determining the foot of the slope as well as allow a claim to be optimized (by defining detailed structure in the bathymetry that allows a claim to be drawn from promontory to promontory). It is thus with either a single-beam echo sounder or, preferably, a multi-beam echo sounder that two of the key elements of a claim (the foot of the slope and the 2,500 m bathymetric contour) are established.

Sediment Thickness

The thickness of sediment under the continental margin (needed to determine the Gardiner Line) can be measured directly through drilling (a rare and extremely expensive venture) or measured remotely through seismic techniques. The most commonly used seismic technique is multichannel seismic reflection, a technique whereby low-frequency seismic sources (typically an array of "airguns" – devices that use compressed air to create a sonic pulse) are fired behind a vessel. The vessel also tows one or more long (often up to 10 km) "streamers," which are devices that record the seismic energy as it bounces off the seafloor and from layers beneath the seafloor. In this fashion a two-dimensional picture of the structure beneath the seafloor is generated as the vessel moves along. With proper processing, the travel times measured by the seismic system can be converted to measurements of depth and the depth of sediment beneath the margin can be determined. This technique is also the primary tool used in the exploration for hydrocarbons.

Gravity and Magnetics

Information about the composition of the earth's crust, its origin and its history can be extracted from measuring its gravity and magnetic signature. While this information is not explicitly required for a claim under UNCLOS Article 76, it can provide important supplemental information in support of a claim. The strength and orientation of both the earth's gravity and magnetic fields are measured with relatively small sensors that can be flown on satellites or aircraft and carried by vessels at sea. These instruments measure the signature of the earth's crust at and below the seafloor; typically, the closer the instrument is to the seafloor, the greater the sensitivity and resolution.

THE CHALLENGE OF THE ARCTIC

As outlined previously, the techniques associated with taking the measurements needed to make a claim under UNCLOS Article 76 are well established. Marine scientists have been collecting high-quality bathymetry, seismics, gravity and magnetics data from the world's oceans for years. A problem arises, however, when trying to implement many of these techniques in the often continuously ice-covered waters of the

Arctic. Because magnetic and gravity measurements can be made remotely through the ice from air- or satellite-borne sensors, they pose less of a challenge (at least at air- or space-borne sensor resolution) than the measurements of bathymetry or sediment thickness that need to be made with the sensors in direct contact with the sea surface.

Early efforts to measure the bathymetry and subsurface structure of the Arctic focused around the establishment of ice islands, laboratories established on large pieces of floating ice that drifted with the Arctic icepack. An example of this was Fletcher's Ice Island (T3), which was a base for scientific investigation as it traversed the polar drift between 1962 and 1974. On this slow-moving platform, holes were cut in the ice and instruments were lowered to measure water and seafloor properties. These instruments included echo sounders and seismic profiling equipment. Over the 12 years the T3 was in operation, approximately 30,000 individual depth measurements were made in regions where no depths had been measured before. While 30,000 measurements may seem substantial, a modern multi-beam echo sounder can easily collect more than 100,000 soundings PER HOUR in equivalent water depths. Nonetheless, the dedication of those on the ice islands provided some of our first glimpses of the nature of the deep seafloor in the high Arctic.

In addition to measurements made from ice islands, many thousands of individual spot depth, gravity, and magnetics measurements were made by scientists who traveled to the ice via helicopter or small aircraft. These added approximately 70,000 soundings to the Arctic database – again, less than an hour's worth of modern multi-beam sonar data. With the development of ice-breakers, single-beam echo sounders and seismic equipment have been brought to a number of areas of the Arctic, adding greatly to our database, but still only providing the two-dimensional view offered by single-beam echo sounders. Collection of these data has not been without great difficulty, even on the biggest of the Russian nuclear icebreakers. The process of ice-breaking is inherently noisy, which makes it difficult to make acoustic measurements. Particularly difficult is the collection of seismic data as ice closing behind the vessel makes it very difficult and dangerous to tow the very expensive seismic gear. The Germans, in particular scientists at the Alfred Wegner Institute, have been especially proficient at developing techniques to collect seismic data in the high Arctic, but the amount of data collected remains very limited.

There is a class of vessels that have access to the Arctic Ocean and that are free from the constraints of the ice: nuclear submarines. For many years nuclear submarines operated under the Arctic icecap and collected single-beam bathymetric information. Thanks in large part to the efforts of Capt. George Newton, much of the bathymetric data collected by U.S. submarines is being declassified and made available to the scientific community. In addition, over the past 10 years there have been five U.S. nuclear submarine missions that have been dedicated to collecting scientific data, including bathymetry. On the last two of these missions (known as SCICEX) a specially built swath bathymetric sonar system was mounted on the submarine to collect wide-area coverage bathymetric and sidescan sonar data; gravity and magnetics data were also collected. The SCICEX submarines also carried a high-resolution sub-bottom profiler, but it did not have the seismic systems needed for determining sediment thickness in support of Law of the Sea claim. While there are some issues with the quality of the navigation associated with the submarine data, the potential of the nuclear submarine as a tool for mapping in the high Arctic is tremendous. The problem, of course, is that they are not easy to come by.

Finally, there looms on the horizon a technological development that may be the best hope for the thorough mapping of the seafloor beneath the ice-covered Arctic. The capability of Autonomous Underwater Vehicles (AUVs) is rapidly developing. There are now systems, such as the Hugin system by Kongsberg Simrad, that can carry multi-beam sonars and collect high-resolution bathymetric data autonomously for periods of up to four days. These systems are still in their early days of development, and while several have ventured under the ice for brief periods of time, it is a bit early in their development to send these multimillion dollar platforms under the ice of the high Arctic for routine mapping missions. In the coming years, however, AUVs may offer an excellent and cost effective option for mapping in support of Law of the Sea and other objectives in the ice-covered Arctic.

CURRENT STATE OF KNOWLEDGE

Given the ability to collect air and space-borne gravity and magnetics, supplemented by spot measurements as well as shipboard and submarine derived data, compilations of Arctic gravity and magnetics have begun to reveal the nature of Arctic crust and tectonic processes (see Appendix, Figure 2 – GSC Magnetic compilation, Verhoef et al.). In particular, the magnetic signature has clearly revealed areas where seafloor spreading has emplaced oceanic crust, though the nature of the crustal type in some areas of the Arctic remains controversial and very relevant to potential UNCLOS Article 76 claims in the Arctic. The least constrained of the primary data sets required for an UNCLOS Article 76 claim (bathymetry and sediment thickness) is the sediment thickness data. As mentioned earlier, collecting seismic data in the ice-covered Arctic is a difficult task and there are very limited amounts of seismic (and thus sediment thickness) data available. The Canadians have produced a very general compilation of sediment thickness (Figure 2, Jackson and Oakey, 1990). Based on hydrocarbon exploration work and a general understanding of the depositional processes at work in the Arctic, it is clear that many of the marginal areas contain thick accumulations of sediment that will certainly be conducive to a Gardiner Line-based claim. It is unclear what data requirements the Committee on the Limits of the Continental Shelf will make for Arctic claims, but it is likely that in some areas, more seismic data will need to be collected.

With respect to the most fundamental data needed for an UNCLOS claim, the recent IBCAO (International Chart of the Arctic Ocean) compilation of the bathymetry of the Arctic has advanced greatly our understanding of Arctic bathymetry (see Appendix, Figure 3, Jakobsson et al.). The IBCAO project, a prime example of international cooperation, compiled data from numerous sources (including recently unclassified data from U.S. nuclear submarines) into a gridded product that offers a depth value every minute, which is equivalent to one mile, for the region above 64 degrees north. While this compilation represents a quantum leap in our knowledge of Arctic bathymetry, the question remains as to whether this data would be useful for resource exploration or for making a claim under UNCLOS Article 76.

MULTI-BEAM MAPPING ON THE CHUKCHI CAP

The Center for Coastal and Ocean Mapping at the University of New Hampshire recently did a study of current U.S. data holdings with respect to a potential claim under UNCLOS Article 76 (Mayer et al., 2002). As part of this study, several areas were identified where the collection of additional data would be particularly useful for making a claim. Most prominent among these was the area of the Chukchi Cap and Northwind Ridge, broad submarine features extending from the Bering Sea and the North Slope of the Alaskan margin. The existing data base from IBCAO and other sources indicated that there was potential for a large extended claim under UNCLOS Article 76 and we set out to test the viability of using modern bathymetric mapping tools on an icebreaker to directly address the UNCLOS Article 76 data needs.

Existing data indicates that in the southern Canada Basin the sediment thickness ranges between 6.5 km in water depths greater than 3,800 m and 11 km where the water depths are 2,000 m (May and Grantz, 1990). The sediment compilation by Jackson and Oakey (1990) further indicates a sediment thickness in the order of 6 km in the northern Canada Basin. This large accumulation of sediment should make it possible to extend a potential U.S. claim based on the sediment thickness (Gardner Line) criterion out to the limit lines (350 nm or the 2,500m contour + 100 nm) as defined in UNCLOS Article 76.

Looking at the existing IBCAO bathymetry and the shape of the coastline of Alaska (defining the territorial baselines for a U.S. claim), it is clear that north of the North Slope of Alaska to a latitude of between 73 and 74 degrees N, the most advantageous cutoff line for a U.S. claim would be the 350 nm line Beyond about 74 degrees north, however, where the Chuckchi Cap and Northwind Ridge form a north-south oriented promontory, the most advantageous cutoff line would become the 2,500m +`100 nm line (see Appendix, Figure 4 from Mayer et al., 2002). Thus to make a claim based on the sediment thickness in this region we would need to: 1) have collected the appropriate seismic data; 2) defined the foot of the slope; and 3) defined the 2,500 m contour.

We began our exploration of the feasibility of using an icebreaker for collecting Law of the Sea relevant data in the summer of 2003, on board the U.S. Coast Guard Icebreaker HEALY. The HEALY, is a 470-foot icebreaker equipped with a Seabeam 2112 multi-beam sonar. For this initial feasibility exercise we chose the 2,500 m contour as our target and attempted to define, follow and map the contour, in real time through 8/9 ice cover. The experiment proved quite successful, and although breaking ice certainly degrades the quality of the multi-beam data that we collected, we were able to map the contour even as it deviated significantly from the existing maps (see Appendix, Figure 5, the Chukchi Cap Map). The newly plotted location of the 2,500 m contour deviates from the existing charts by as much as 40 kms in some places, having a significant effect on a potential claim under UNCLOS Article 76.

An even more poignant example of the need for more survey work in the Arctic is the discovery during our short cruise of an unmapped seamount (now officially named "HEALY Seamount") that rose from 4,000 m to less than 900 m in an area where the existing charts showed nothing more than a single 2,500 m contour (see Annes, Figure 6). The fact that the 2,500 m contour was, in some areas, significantly different from that depicted in the earlier charts, and that features as large as HEALY Seamount were totally unmapped supports the need for further survey work before an accurate claim can be made. In the coming years we plan on continuing our mapping efforts, finishing mapping the location of the 2,500 m contour followed by multi-beam sonar mapping of the foot of the slope. We will also have to address the need for the collection of seismic data.

CLOSING REMARKS

The harsh environmental conditions and ice-covered waters of the Arctic pose great challenges to those charged with mapping the region. There has been great progress over the years in the compilation of gravity, magnetics, seismic and bathymetric data in the Arctic, but even the most recent compilations are probably not adequate for making a claim under UNCLOS Article 76. While the technologies used to collect and process the data needed to support a continental shelf claim under UNCLOS Article 76 are well defined and developed, their implementation in the ice-covered areas of the Arctic is extremely challenging. Complete coverage multi-beam sonar data is the desired bathymetric product for making a claim. Swath-sonar equipped nuclear submarines and AUVs may provide an elegant solution, but these assets are not readily available at this time. A multi-beam, sonar-equipped ice-breaker proved a viable way of collecting the bathymetric data needed to support a claim, but progress is slow. Unlike the vagaries of maritime law, there is only one bathymetric truth in the Arctic and it should be our goal to seek this truth. Given the challenges before us it seems reasonable that we should seek a collaborative international effort to better map and understand the Arctic.

References

Jackson, H.R., and Oakey, G.N., 1990, Sedimentary thickness map of the Arctic Ocean, Plate 5 in Grantz, A., Johnson, L., and Sweeney, J. F. (Eds.), The Arctic Ocean Region, *The Geology of North America* DNAG v. L.

Jakobsson, M., N.Z. Cherkis, J. Woodward, R. Macnab, and B. Coakley. 2000 New grid of Arctic bathymetry aids scientists and mapmakers; *Eos*,Transactions, American Geophysical Union, v.81, no.9, p. 89,93,96.

May, S.D., Grantz, 1990, A., Sediment Thickness in the Southern Canada Basin, *Marine Geology*, v. 93, p. 331-347.

Mayer, L.A., Jakobsson, M., Armstrong, Andrew, 2002, The Compilation and Analysis of Data Relevant to a U.S. Claim Under United Nations Law of the Sea Article 76, <http://www.ccom.unh.edu/>

Appendix

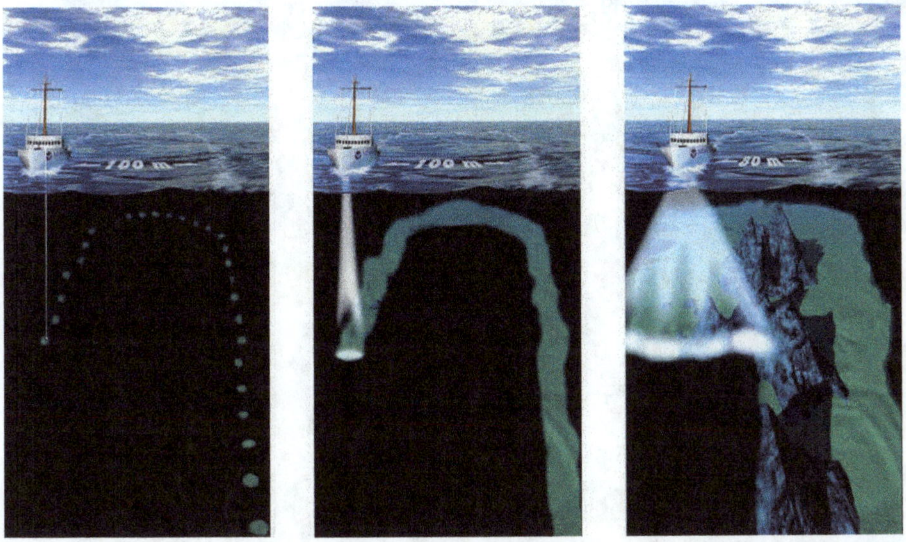

Figure 1: Comparison of lead line, single-beam and multi-beam sonar. Figure courtesy of NOAA.

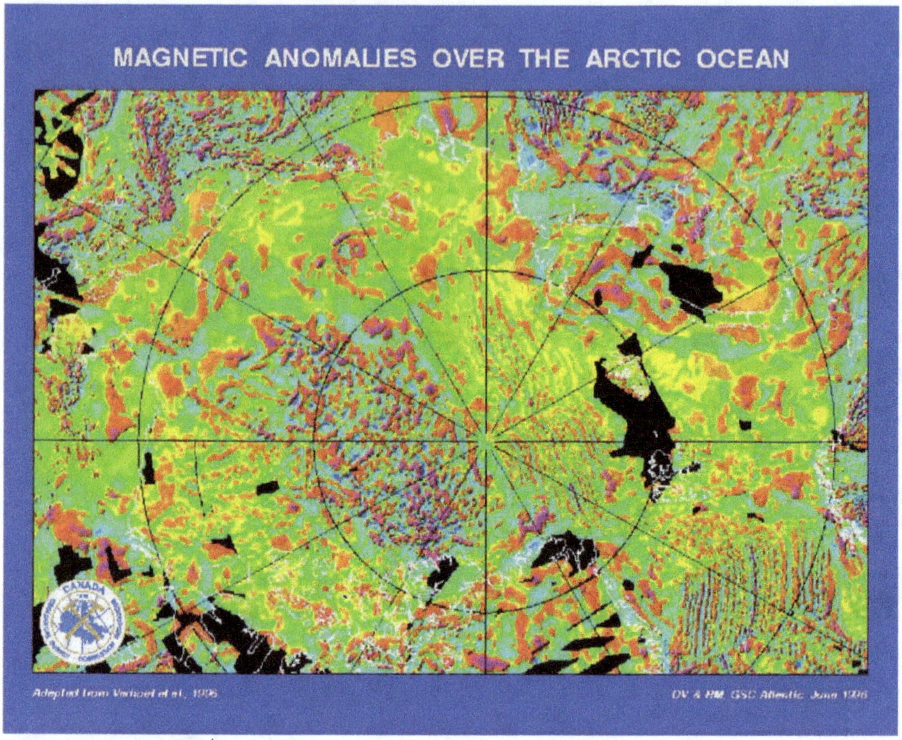

Figure 2: Magnetic compilation of the Arctic produced by the Geological Survey of Canada, from Verhoef et al., 1996.

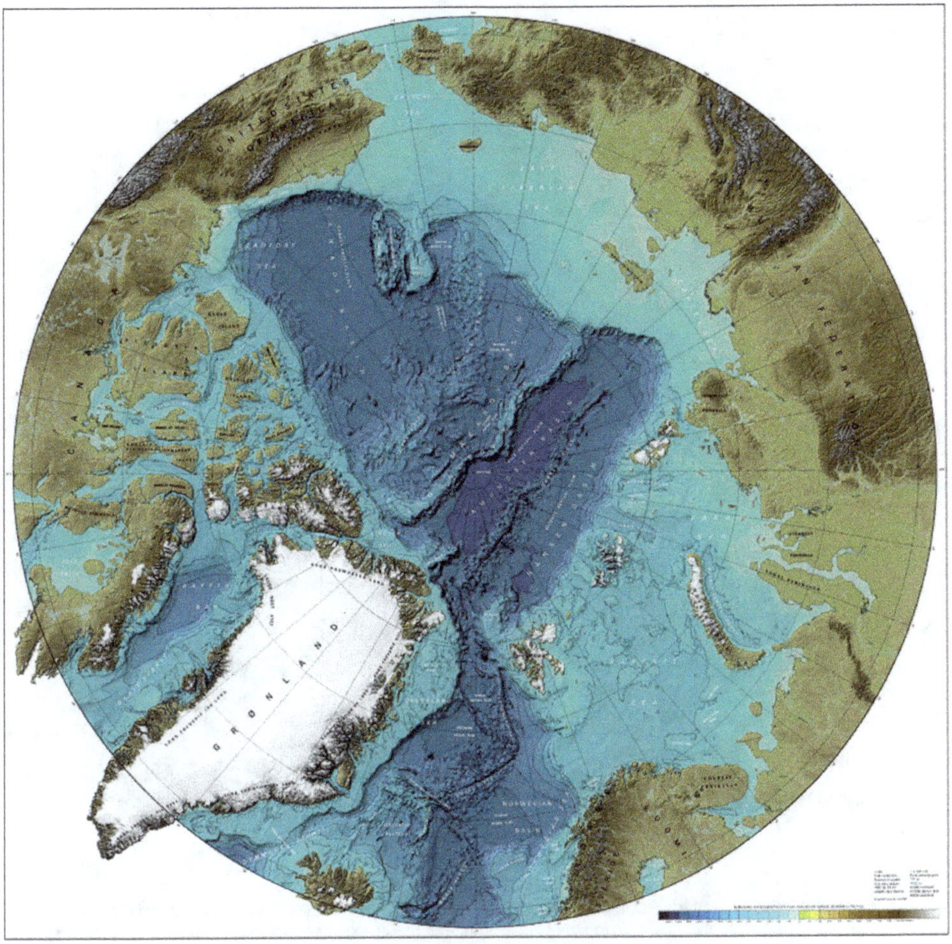

Figure 3: New IBCAO map of Arctic bathymetry (Jakobsson, et al., 2000).

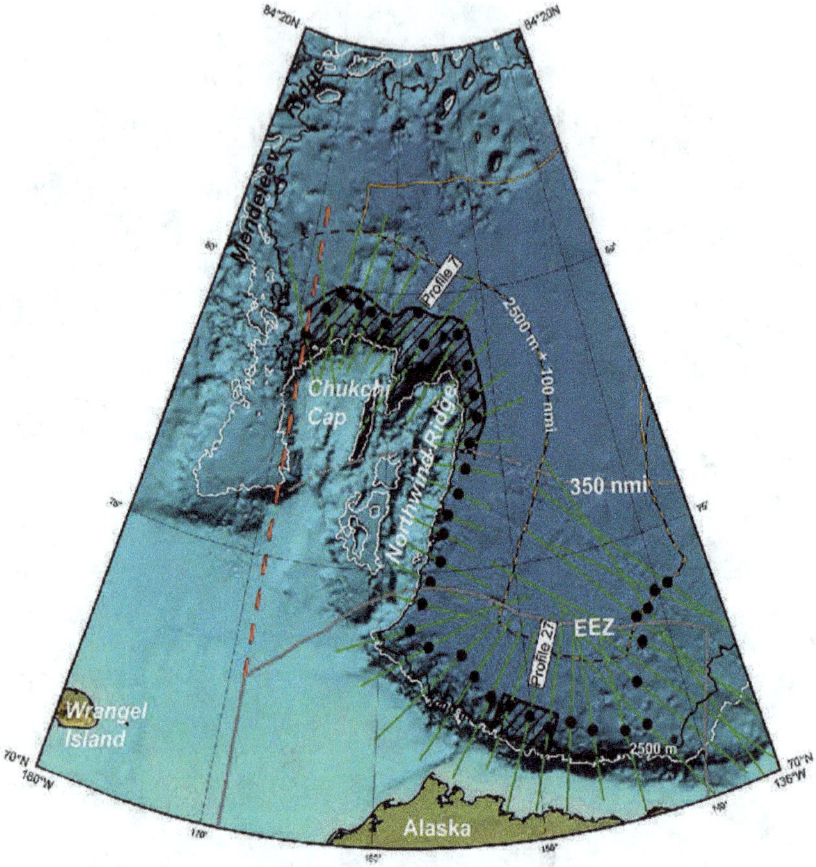

5.10B. Bathymetry from IBCAO in detailed area ARC, drawn bathymetric profiles, and possible locations of the FOS. Labeled profile is shown in figure 5.11. Note that the orange line, which represents the 2500 m + 100 nm, makes use of the 2500 m contour of the Alpha-Mendeleev Ridge as well as the Canadian shelf.

Figure 4: Chukchi Borderland cutoff lines. Due north of the Alaskan coast, the 350 nm line is the most advantageous cutoff line, but east of the Northwind Ridge, the 2,500 m isobath + 100 nm line becomes the most advantageous cutoff.

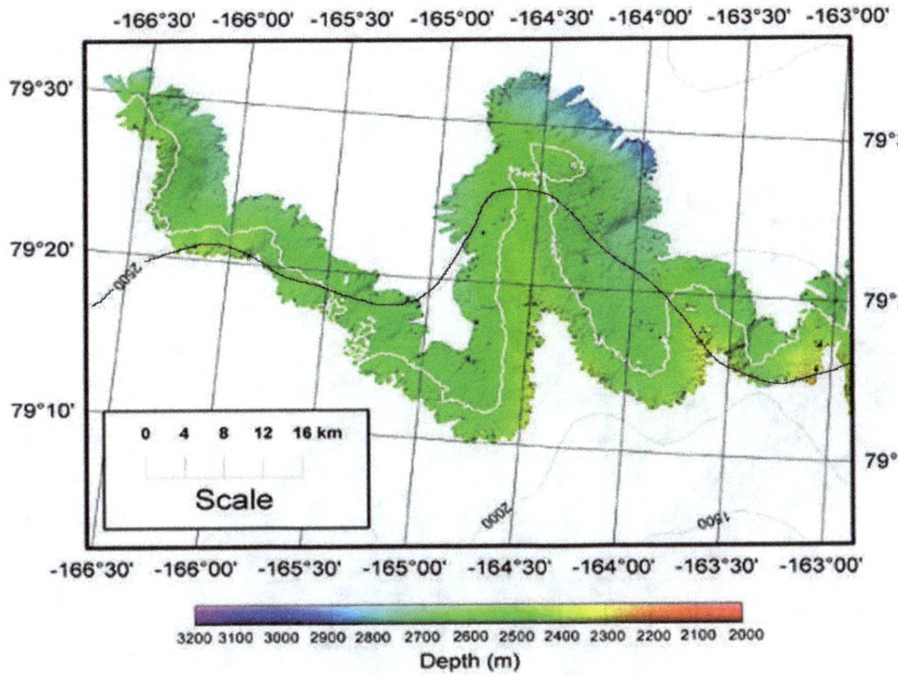

Figure 5: Comparison of the 2,500 m contour as depicted on the IBCAO chart (black line) with the 2,500 m contour as mapped by multi-beam sonar.

Figure 6: Three dimensional perspective of the unmapped Healy Seamount, rising from depths of 4,000 m to approximately 900 m.

PANEL IV:
LEGAL, ENVIRONMENTAL AND TECHNICAL ASPECTS OF ENERGY TRANSPORTATION

The Turkish Straits, Oil Transportation and Turkish Policy

Nilüfer Oral[*]

The Turkish Straits have maintained an influential role in regional and international politics for centuries. Indeed, the matter of the Turkish Straits was the very heart of the "Eastern question" that defined much of Western European politics during the 19th century and early 20th century. A central theme of that time was Czarist ambition to control the Straits and British policy to prevent it. During the first half of the 20th century the Turkish Straits weaved in and out of the dangerous waters of two world wars and the Cold War. During the second half of the 20th century the role of the Turkish Straits in regional and international politics appeared to have diminished but then re-emerged with the discovery of significant reserves of oil in the post-Soviet CIS countries.[1] This once again placed the Turkish Straits under the limelight of regional and international politics.

While at one time the importance of the Turkish Straits was for war, today it is for oil. The increase in upstream oil production in the Caspian and Caucasus together with the need to export to Western markets has placed pressure on this narrow stretch of water serving as the sole water link between the Black Sea and the Mediterranean Sea. The amount of oil and oil products transported from the Black Sea through the Turkish Straits has increased from 60 MTA in 1996 to nearly 150 MTA in 2003. The number of tankers has likewise increased from just over 4,200 in 1996 to more than 8,000 in 2003.[2] The risk of an environmental incident is quite real, as demonstrated by numerous accidents involving oil tankers in the past. The most serious of these was the 1979 *Independenta/Evriali* collision, which resulted in an oil spill of 94,000 tons of oil, and the 1994 *Nassia/Shipbroker* collision, which resulted in an oil spill of 20,000 tons.

In addition to the environmental pressures and dangers placed upon the Turkish Straits there is political pressure, as the Black Sea region becomes a new oil region. As a strategic commodity, the role of oil goes

---

* The views expressed in this paper belong solely to the author and do not represent the views of any institution or government. Dr. Oral is Assistant Director of the Istanbul Bilgi Marine Law and Policy Research Center.

beyond mere economics occupying a pivotal place in the balance of power between states. The Black Sea region is no exception. The collapse of the former USSR, a superpower, has resulted in a smaller and much weaker Russian Federation as its replacement. The USSR, whose oil production was once the greatest in the world, exporting nearly 11 million barrels a day, was forced into giving up some of the richest oil and natural gas reserves when countries such as Azerbaijan, Kazakhstan, Turkmenistan gained independence and formed part of the Newly Independent States (NIS) and Commonwealth of Independent States (CIS). For Russia, control over these rich energy resources became of primary importance.

When Turkey decided to take measures to enhance navigational safety and protection of the marine environment in the Turkish Straits during the early 1990s, Russia viewed this as a possible threat to its navigational rights in the Turkish Straits, and also as a possible attempt to modify the provisions of the 1936 Montreux Convention for the Turkish Straits. However, neither was true. Turkey had been studying measures to take in the Turkish Straits as a result of increased maritime incidents.[3] Furthermore, the measures taken were viewed as being consistent with the navigational rights established by the 1936 Montreux Treaty and international law.

THE 1936 MONTREUX CONVENTION AND NAVIGATIONAL RIGHTS IN THE TURKISH STRAITS

The legal history of the Turkish Straits is rich with treaties.[4] However, the only treaty that "internationalized" the Turkish Straits was the 1923 Lausanne Peace Treaty provisions on the Turkish Straits. The 1923 Lausanne Treaty was concluded between the European powers that had unsuccessfully invaded the Ottoman Empire after its defeat in World War I and the newly established Turkish Republic.[5] The Lausanne Treaty marked the international recognition of the Republic of Turkey, defining its borders, international obligations and rights. It also included a separate part regulating navigation of vessels, particularly warships, through the Turkish Straits. However, its provisions on the Turkish Straits resulted in imposing limitations on Turkish sovereignty over the Straits by demilitarizing the Straits[6] and by creating an international commission responsible for their administration.

Thirteen years later, in 1936, Turkey successfully sought the renegotiation of the 1923 Lausanne Treaty provisions on the Turkish

Straits. The 1936 Montreux Convention Regarding the Regime of the Straits ("1936 Montreux Convention") restored Turkey's full sovereignty over the Straits by abolishing the demilitarized status of the Turkish Straits and the international guarantee as well as the international commission.[7] The new Convention also included detailed provisions on the regime of passage of warships into and out of the Black Sea through the Turkish Straits.

The return of Turkey's full sovereignty over the Straits under the Montreux Convention was unconditional. No provision in the text itself or in the proceedings of the conference imposed either directly or indirectly any condition of sovereignty over the Straits. In fact, one of the significant aspects of the Convention, as noted by Rozakis and Stagos, was that Turkey "under the Montreux Convention, [was] not just the sovereign of a territorial region with international obligations to fulfill, but also the regulating factor in the functioning of the Convention, with leeway allowed by an international agreement to implement the regulations. Turkey's jurisdiction under the Montreux Convention is a central element in the evaluation of its role in international affairs."[8]

Furthermore, there is little factual support for the view held by some Russian writers that the provisions regulating passage of commercial vessels were the "most important." To the contrary, the primary concern of the Contracting Parties in July 1936 was not the right of freedom of navigation for commercial vessels, but rather the rights of passage into and out of the Black Sea through the Turkish Straits. The emphasis of the Montreux Convention was on war and warships. The Montreux Convention, with a total of 29 Articles, three annexes and one protocol, devoted Articles 2 through 7 to the passage of merchant ships. The bulk of the Convention (Articles 8 through 22) regulated the passage of war vessels. Likewise, the bulk of the discussions during the month-long conference focused on the passage of warships.

This did not, however, mean that the provisions concerning freedom of navigation for commercial vessels were of no importance. The principle of freedom of passage was recognized in Articles 1 and 2 of the Montreux Convention. However, the meaning to be ascribed to "freedom of passage" was qualified by the text of the Convention itself, as well as by statements made during its negotiation. For example, one scholar concluded in 1938 that the Montreux Convention had ushered in a new passage regime for the Turkish Straits:

> . . . Freedom of navigation, the only surviving provision of the Lausanne Convention also underwent profound changes: the waters of the Straits no longer formed an open corridor at all times. . . . for the complete freedom of navigation of all foreign vessels. The Montreux Convention established limitations for both the passage of merchant vessels as well as war vessels.[9]

The limiting character of the Montreux Convention on the principle of freedom of passage was reiterated by other eminent international legal scholars such as Brüel,[10] Rozakis and Stagos.[11] Furthermore, during the Montreux conference itself the Turkish delegation clearly stated that the principle of freedom of passage for the Straits was not to be that of the high seas freedom. In fact, Mr. Menemencioğlu, the head of the Turkish delegation, expressly admonished the other participants not to come back and approach Turkey in the future claiming that "ships (could) navigate in the Straits as if they were on the high seas."[12] According to the Turkish delegation, the meaning to be given to the "freedom of passage" of commercial vessels would be subject to Turkey's authority to "police navigation"[13] to ensure that passage is "inoffensive" or "innocent." When introducing the first set of Turkish Draft Articles, Mr. Menemencioğlu specifically declared that the passage of vessels through the Straits was always to be innocent ("inoffensive") and non-belligerent.[14] Turkey also made clear its intention, notwithstanding the text of the Convention, to retain administrative (police) authority and judicial authority in the Turkish Straits as such powers fell outside the scope of the Convention. None of the Representatives at the Conference contested these declarations.

In summary, the express restrictions placed on the passage of merchant vessels by the text of the Convention itself, together with the official statements made during the Conference by Turkish Representatives, defined the parameters of "freedom of passage and navigation" for merchant vessels. Turkey had expressly reserved its authority to protect the Straits, which included safety of navigation.[15] The Montreux Convention thus created a regime of passage for merchant vessels that incorporated the general principles of an innocent passage regime but, in addition, included unique conditions of passage.

It is against this historical and legal background that the subsequent enactment of the 1994 Maritime Traffic Regulations for the Turkish

Straits (1994 Regulations)[16] and the 1998 Turkish Straits Maritime Traffic Scheme Regulations (1998 Regulations)[17] must be analyzed.

THE 1994/1998 TURKISH STRAITS REGULATIONS

With the exception of the occasional controversy over the passage of warships, international interest in the legal nuances of the Montreux Convention was, until recently, negligible.[18] However, in Turkey, particularly in the aftermath of the disastrous and tragic 1979 *Independenta* accident, regulation of commercial traffic through the Straits, particularly tankers transporting hazardous cargo, has begun to attract increased legal attention. In 1979 Professor Toluner published her landmark article on commercial passage regime, wherein she outlined the legal foundation for the application of the regime of innocent passage under Articles 1 and 2 of the Montreux Convention.[19] Other Turkish scholars have since supported her position.

Well before the dissolution of the USSR and the discovery of rich oil reserves in the former USSR States, now referred to as the CIS, the Association of Turkish Ocean Going Masters in 1987 prepared and submitted to the Ministry of Transportation a report and proposal for the creation of a traffic separation scheme in the Turkish Straits.[20] In 1990, a Commission was established to conduct a detailed study of the matter of safety of navigation through the Straits. The Commission concluded that a traffic separation scheme was necessary to ensure the safety of the Straits as well as the bordering coastal area together with new regulations to replace the 1982 Istanbul Port Regulations. The Commission drafted the Regulations, which were ultimately adopted by the government and enacted as the 1994 Maritime Regulations for the Turkish Straits.[21]

The political question for Turkey was whether or not to submit the proposed traffic separation scheme (TSS) to the International Maritime Organization (IMO). Although Turkey was not under any obligation to do so, either under the 1936 Montreux Convention or any other international Convention, the decision was made to seek international support. On March 26, 1993, Turkey presented an information paper to the 62nd Maritime Safety Committee (MSC) of the IMO on the existing maritime traffic problems in the Turkish Straits and a plan to introduce a TSS in the Turkish Straits in an effort to increase safety of navigation and protect the marine environment. Turkey also advised of its intention to enact new maritime Regulations.[22]

Almost one year to the day Turkey presented MSC 62/INF.10, the *Nassia/Shipbroker* accident shook Istanbul.[23] It sounded an alarm bell for the urgent need for a new traffic management system in the Straits. The IMO was quick to respond and at MSC 63 adopted the Turkish TSS and associated *Rules and Recommendation on Navigation through the Straits of Istanbul, the Strait of Çanakkale and the Marmara Sea,* subsequently confirmed by the Nineteenth General Assembly meeting (SN/Circ.166).[24]

This decision to seek the support of the IMO, an international organization that at the time had nearly 160 members, would certainly provide evidence against Turkey's intention to act in a *unilateral* fashion in the Turkish Straits. To the contrary, Turkey sought multilateral support. However, the issue of the new traffic regulations in the Turkish Straits proved extremely controversial, with Russia playing the role of the lead antagonist. Furthermore, after a series of bilateral and multilateral meetings with various member governments, Turkey reassessed some of the controversial provisions in the 1994 Regulations and revised them.[25] However, the main provisions establishing a Traffic Separation Scheme, a ship reporting system and application of international standards established by SOLAS and COLREG remained.[26]

Russia's primary concern appeared to be that Turkey was seeking to alter the 1936 Montreux Convention. Russia specifically called upon the IMO Legal Committee to render an opinion by submitting a lengthy paper to the Legal Committee entitled "Unconformity [sic] of the Turkish Regulations for Traffic Order in the Area of the Straits and the Sea of Marmara to the Rules and Recommendations adopted by the Maritime Safety Committee." The Russian Federation was essentially lodging a complaint against Turkey and requesting the Legal Committee to render a judgment,[27] a highly questionable action as the IMO not only lacks such authority under its own Constitution but has no jurisdiction over the Turkish Straits under the 1936 Montreux Convention. It also resulted in increasing political tension in the IMO, with Turkey assuming an increasingly defensive posture. Turkish defensiveness was, perhaps, most provoked by the Russian proposal for the creation of an international commission to oversee the operation of the Rules and Recommendations in the Turkish Straits, raising the specter of the Lausanne international commission.[28]

Between 1993 and 1999, the Turkish Straits issue became a source of conflict and a potential failure for the IMO.[29] The United States head of delegation perhaps best summarized the problem in his statement at MSC

71: ". . . since 1994, we believe that discussions that have taken place in this committee have focused on everything but the safety issues of navigation through the Turkish Straits."[30] During the same MSC meeting a Working Group on the Turkish Straits was formed and instructed to examine the pros and cons of the application of Rule 9 and Rule 10 of COLREG,[31] "taking into account the level of safety and protection of the marine environment which (had) been achieved under the existing IMO-adopted system. . . ." These terms were further narrowed to taking into account "the human element," thereby excluding all other factors, including economic and, of course, legal and political. Simply stated, the Working Group was instructed to assess the traffic situation in the Turkish Straits before the 1994/1998 Regulations were enacted and afterwards.

In light of the significant reduction of maritime accidents in the Turkish Straits since the implementation of the national Regulations and the IMO-adopted TSS and Associated Rules and Recommendations in 1994, the MSC 71 concluded that the maritime traffic system in the Turkish Straits had proven to be effective and successful. The MSC 71 recommended that discussions of the Turkish Straits should be discontinued.[32] This was confirmed by the Twenty-first Assembly with the continuing reservation of the Russian Federation. Most recently, Russia submitted a paper to the MSC 78 complaining of delays in passage through the Turkish Straits.[33] Turkey in turn submitted a paper providing statistical information on the operation of the new VTS system, claiming that "…the TSVTS has increased the efficiency of the traffic organization without compromising on the safety rules and regulations."[34]

However, the assertion that the IMO decision to discontinue discussions on the Turkish Straits was based on NATO's need for Turkish support in the war against Yugoslavia is interesting indeed, but not supported by the facts. First of all, Turkey was at the time and has continued to be a NATO member with an existing policy against Serbian attacks on Bosnian Moslems. Clearly, no additional cajoling would have been necessary by NATO or the Clinton Administration to gain Turkish support against Serbian aggression.

Moreover, by 2001 the Turkish Straits was no longer an issue in the IMO. Thus, any Russian view that Turkish administration of the Straits was a matter for negotiation toward a possible war in Iraq appears to be a stretch of the political imagination. Likewise, assigning purely political motives to Turkish measures to protect the Straits against potential maritime disasters, such as those experienced in the past, is also not borne by the facts. The same allegations that the 1994 Regulations were adopted to force the construction of the Baku-Tblisi-Ceyhan pipeline continue 10 years later, even though the pipeline is scheduled to be completed and running by the year 2005.

However, the role of the *Erika* and *Prestige* oil spills did highlight the environmental consequences of an oil spill. In response, several European countries, such as Spain and Portugal, imposed unilateral measures, such as banning single-hull oil tankers from their waters. Turkey, on the other hand, did not ban the passage of tankers but in 2002, the Undersecretariat for Maritime Affairs issued a set of "administrative guidelines" that included daytime passage for large tankers carrying dangerous/hazardous cargo. By 2002 the number of tankers carrying oil and oil products had increased to nearly 7,500. The purpose of the guidelines was to reduce the risk of an accident by managing, not preventing, passage of tankers carrying hazardous cargo. Nevertheless, at times the efforts to balance safety concerns with commercial concerns created discord.

Oil Transportation, the Black Sea and Regional Cooperation

Without doubt, the Turkish Straits serve as a vital transport route for ships traveling between the Black Sea and the Mediterranean Sea. However, the Black Sea itself also serves as an important link in the transport chain of oil from the region to the West. There are several major oil terminals in the Black Sea, and as of 2003 some 150 MTA of oil and oil products was transported. In many ways, the environmental risks facing the Turkish Straits due to the increase in the amount of oil and oil products being transported apply equally to the Black Sea, a body of water that until recently was diagnosed as being moribund.[35]

The Black Sea has been the subject of intense environmental attention since the break up of the former USSR Under the auspices of the United Nations Environmental Programme, a Black Sea Regional Sea program has been in effect since 1991.[36] Once considered to be a dying sea with questionable prospects of recovery, the Black Sea now shows evidence of recovery. This positive trend could, however, be upset should there be an accident involving an oil spill, including a bunker spill.

Although vessel source pollution in the Black Sea accounts for approximately 10 percent of marine pollution, the increase in the amount of oil and the number of tankers transiting the Black Sea merits closer attention to ensure safety of navigation and protection of the marine environment. For this reason, regional cooperation is of the utmost importance. The Regional Seas program is based upon regional cooperation and can only be successful with the cooperative participation of the region. The duty of States to cooperate is a norm of international law and has been emphasized in many instruments, including the 1982 United Nations Convention on the Law of the Sea.

Turkey seeks to reduce the risks associated with the transport of dangerous and hazardous cargo through the Turkish Straits. However, this objective should apply equally to the Black Sea as well. For this reason, all stakeholders should endeavor to promote greater cooperation in a regional context in a manner consistent with international law. The Black Sea offers various means to achieve this, such as the Regional Sea Programme and the Black Sea Memorandum of Understanding for Port State Control.[37] The emerging norms of international law, such as the precautionary principle and "the polluter pays" principle, coupled with the general duty under international law to protect and preserve the marine environment, imposes a duty on the coastal States as well as users of the Black Sea to take the necessary measures to protect against environmental damage resulting from an oil spill.

Conclusion

Turkish policy has been to maintain the integrity of the 1936 Montreux Convention. The 1994 Regulations, as replaced by the 1998 Regulations, were designed to be consistent with the view that the Montreux Convention established a unique regime granting innocent passage rights to all commercial vessels. The confusion lies in the view that the Convention created a "high seas" or "transit passage" regime. However, a careful examination of the text of the Convention, together with the proceedings of the Conference, fails to support this view. As innocent passage rights include the right of the coastal State to adopt laws and regulations in respect of safety of navigation and the regulation of maritime traffic, as well as preservation of the environment and the prevention, reduction and control of pollution,[38] there should be no concern that Turkey has unilaterally amended the Convention. To the contrary, Turkish action in submitting the Turkish Straits Traffic Separation Scheme plan to the IMO in 1993 and continuing to seek a solution to the long-lasting discussions surrounding the maritime regulations show that Turkey adopted a cooperative and multilateral approach.

The problems of oil transport should be looked at through a broader lens and encompass the Black Sea region as a whole. The increase in oil entering the Black Sea necessitates a clear policy of regional cooperation to safeguard both the safe and secure transportation of that oil. The Black Sea itself faces serious environmental pressures, which are being addressed through the UNEP Regional Seas Programme. The safe and secure transportation of oil in the Black Sea will also result in a safer and more secure transportation through the Turkish Straits.

Notes

[1] The oil reserves are deemed to parallel that found in the North Sea. See www.eia.doe.gov

[2] Statistics provided by the General Directorate for Coastal Safety and Salvage.

[3] The Turkish Straits had experienced over 400 maritime accidents during a 50-year period between 1948-1998. Information obtained from Undersecretariat of Maritime Affairs.

[4] See Gürkan, İ., "Turkish Straits Regime in an Historical Perspective (1453-1936) "in *Turkish Straits New Problems New Solutions* (Istanbul 1996) 165-221.

[5] The Turkish Republic was officially proclaimed on 29 October 1923 bringing to end the 600 hundred year rule of the Ottoman dynasty.

[6] Turkish security was provided by an international guarantee.

[7] Rozakis, C.L. and Stagos, P. N., The Turkish Straits (Martinhus Nijhoff, Dordrecht 1987), p. 115.

[8] *Ibid.* at p. 120.

[9] Djonker, M.C., *Le Bosphore et Les Dardanelles Les Conventions des Detroits de Lausanne (1923) et Montreux (1936)* (1938 These de Doctorat), p. 146. (Transl. by N.Oral).

[10] Brüel, E. *International Straits A Treatise on International Law* (1947), 393. Brüel observed that the "apparent liberality of the rules of passage contained in Article 2 is considerably modified by the fact that shipping is subject to sanitary control under Article 3 and the dues set out in Annex I." P. 409.

[11] *Ibid.* at pp. 106-109. Brüel was critical of the Montreux Convention because of the limitations placed on freedom of navigation.

[12] Meray, S.L., Olcay, O., *Montreux Boğazlar Konferansı*, Tutanaklar, Belegeler (Istanbul: 1976) p. 164.

[13] *Le Bosphore et Les Dardanelles Les Conventions des Detroits de Lausanne Ibid* N.Oral).

[14] *Montreux Boğazlar Konferansı, Ibid.* p. 42.

[15] *See* Toluner, S., "Rights and Duties of Turkey Regarding Ships in Transit Through the Straits," *Boğazlardan geçiş Güvenliği ve Montreux Sözleşmesi* (İstanbul 1994) p. 73.

[16] Came into effect 1 July 1994, repealed 6 November 1998.

[17] Went into effect 6 November 1998.

[18] International interest was restricted to scholarly studies such as DeLuca, A.R., *Great Power Rivalry at the Turkish Straits: the Montreux Conference and Convention of 1936* (New York 1981) pp. 115-117 and Rozakis and Stagos, *Ibid.*

[19] Toluner, S., "Boğazlardan Geçişi Düzenleme ve 'Montreux'" *Milliyet,* 3 December 1979, see also, Toluner, "Rights and Duties of Turkey Regarding Merchant Vessels Passing through the Straits" *Turkish Straits New Problems New Solutions* (Ed I. Soysal) (Istanbul 1995) p. 27.

[20] From the private archives of Captain Gündüz Aybay.

[21] The Regulations went into effect on July 1, 1994.

[22] MSC 62/INF.10.

[23] March 13, 1994.

[24] IMO Doc. MSC 63/WP.5/Add.1; MSC 63/WP.17. IMO doc. Res. A/827.19

[25] The 1998 Regulations went into effect 6 November 1998.

[26] For a more detailed discussion of the Turkish Straits Regulations see, Oral, N., "Oil and Water: Caspian Oil and Transportation Challenges Facing the Turkish Straits" in *Current Marine Environmental Issues and the International Tribunal for the Law of the Sea (Eds. Nordquist, M.H. and Moore, J.N.)* (Marinus Nijhoff Hague 2001) 329; Ünlü, N., *The Legal Regime of the Turkish Straits* (Martinus Nijhoff Hague 2002).

[27] The Russian document once again put forth the view that the national Regulations put into effect by Turkey on July 1, 1994, were in contradiction with the 1936 Montreux Convention and international law in general, including the 1982 UN Convention on the Law of the Sea and not in conformity with the IMO Rules and Regulations as adopted by the Sixty-third MSC and issued as SN/Circ. 166. The Russian paper requested the Legal Committee to give a direct opinion as to whether the Turkish Regulations were in conformity with IMO Rules and Recommendations, and specifically, the Russian Federation objected to those provisions of the national Rules, which:

> envisaged in some cases de facto the imposition of an authorization procedure for the passage, other unreasonable restrictions for navigation including its suspension under certain circumstances or even a ban for certain classes of ships for passing through the Straits.

See IMO Doc. LEG 71/12. The Legal Committee concluded that the national Regulations were not in compliance with international law and IMO Rules and Recommendations but that the matter should be resolved by the MSC.

[28] IMO Doc. MSC 67/7/8

[29] For detailed analysis of the Turkish Straits at the IMO see see Unlu, *The Legal Regime of the Turkish Straits Ibid.*, Oral, "Oil and Water Caspian Oil and Transportation Challenges Facing the Turkish Straits *Ibid.*; Plant, G., "The Turkish Straits and Tanker Traffic: an Update" *Marine Policy* 24 (2000) 193:214; Dyoulgerov, M., "Navigating the Bosphorus and the Dardanelles: A Test for the International Community," 14 *Intl Journal of Maritime and Coastal Law* 1 (1999); Scharfenberg, A., "Regulating Traffic Flow in the Turkish Straits: A Test for Modern International Law," 10 *Emory International Law Review* 1 (Spring 1996);

[30] Based upon the personal notes of the author.

[31] Rule 9 of COLREG, which was the rule of navigation in force in the Straits of Istanbul before the 1994 Turkish Straits Regulations were enacted, provided for right side passage allowing much discretion to the ship's master. Whereas Rule 10 required ships to navigate within the designated traffic lanes of the TSS allowing virtually no discretion to the ship's master.

[32] IMO Doc. MSC 71/23

[33] IMO Doc. MSC 78/25

[34] IMO Doc. MSC 78/INF.16

[35] See Black *Sea Trans boundary Diagnosis* at http://www.grid.unep.ch/bsein/

[36] The Convention on the Protection of the Black Sea Against Pollution was signed in Bucharest, along with three Protocols in April 1992, and ratified by all six Black Sea countries in 1994. A fourth Protocol, The Black Sea Biodiversity and Landscape Conservation Protocol to the Convention on the Protection of the Black Sea Against Pollution, was signed in 2003. See http://www.blacksea-environment.org/.

[37] Signed in April 2000.

[38] See article 21 of the 1982 LOSC, which enumerates the areas in which the coastal State may adopt laws and regulations relating to innocent passage consistent with the Convention itself as well as other rules of international law.

Cooperation Between Vessel Traffic Services (VTSs) in the Black Sea

Capt. Robert Hofstee[*] *and Capt. Dr. Ozkan Poyraz*[†]

EXECUTIVE SUMMARY

Various instruments are available to governments that wish to protect their environment from negative effects of shipping. The recent introduction of Vessel Traffic Services (VTSs) in the Turkish Straits, the use of AIS in conjunction with VTSs in the Black Sea and Port State Control are issues that need to be investigated in order to establish whether the apparent overall safety of the areas concerned can be improved. Any changes in technical, operational or regulation practices of sub-areas may affect the whole Black Sea area. This paper includes a short description of relevant directives of the European Communities and some European Community Research Projects and the effects they could have at a later stage in the Black Sea area. The introduction of AIS-derived data in a centralized database is described, and a practical approach of cooperation between important parties concerned with safety is advocated, as well as exchange of information via databases, training and twinning.

[*] Consultant & visiting Professor, WMU-Malmö, Sweden, info@hofstee.biz.
[†] Assistant Professor, ITU-MF, Istanbul-Tuzla.

INTRODUCTION

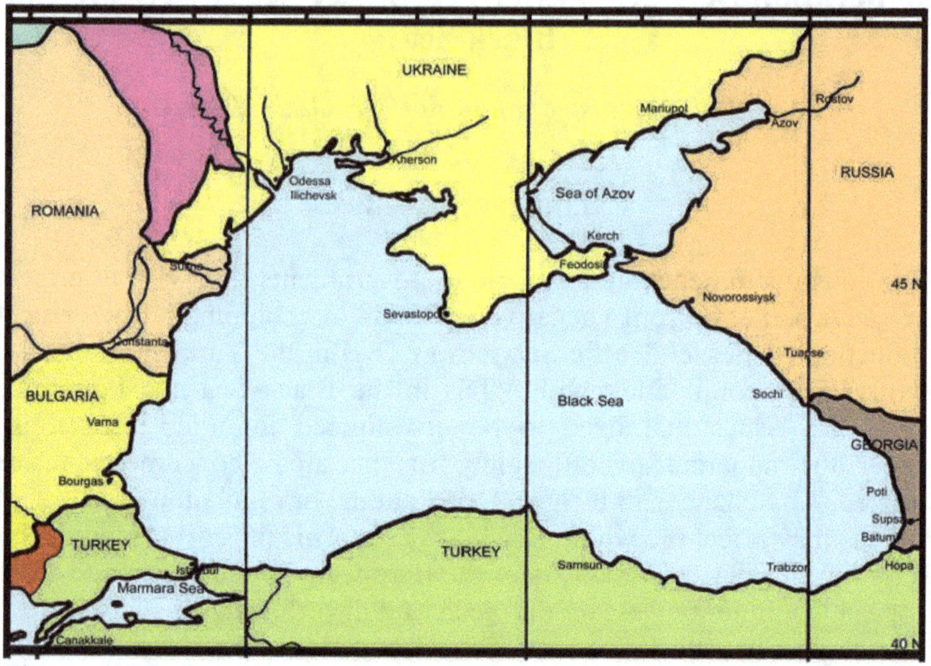

Figure 1: Black Sea region and major ports

Maritime transport in the Black Sea region is of utmost importance to all states bordering the Black Sea. It is necessary for them to ship their products to other parts of the world and to import the commodities they need.

Vessels of various flag States use the connection between the Mediterranean Sea and the Black Sea in order to reach their destinations. The dimensions of these vessels can be quite large compared with the narrow straits they use for transit. At the same time, the training of the various crews on board these vessels is not always of the same high standards, and often the vessels are in poor condition. There are, however, a number of instruments that control the vessels and their crews, mainly international Conventions of IMO[1] and ILO. The actual implementation of these conventions is checked by the flag States and by Port State Control.

Apart from international Conventions and Recommendations, states, as well as ports, have their own regulations for their territory. The Turkish Straits are a special case; they are governed by the Montreux Treaty of 1936 and associated Regulations of the Turkish State.

Some specific regulations that apply to the European Union will be mentioned, though not *in extenso*, at a later stage.

VTSs have been established in ports and approaches all over the Black Sea area. The latest VTS to become operational was the Turkish Straits VTS.[2]

Figure 2: Existing VTS stations and coverage in Black Sea

Pilotage plays an important role with regard to the safety of shipping and the protection of the environment. In most cases pilotage is mandatory, but not in the Straits leading toward the Black Sea area from the Aegean Sea.

The use of IMO-approved Traffic Separation Schemes (TSS) certainly contributes to the overall level of safety, as long as vessels comply with the regulations governing the TSS. IMO-approved reporting schemes assist coastal States and ports in assessing the expected traffic.

The tragic events of September 11, 2001, resulted in the introduction of some new safety schemes and in existing schemes being speeded up.[3] At a European level, the European Maritime Safety Agency[4] (EMSA) should be mentioned. Its main tasks are the following: strengthen the Port

State Control regime; audit the Community-recognised classification societies; develop a common methodology for the investigation of maritime accidents; and establish a Community vessel traffic monitoring and information system.[5] Recently, according to a decision of the European parliament, some extra tasks were added: develop an operational capability in oil pollution response; perform quality control of maritime training in third countries; and perform responsibilities related to monitoring maritime security issues and advise the European Commission accordingly.

CONTROL OF VESSELS

If we wish to make maritime traffic in a specific area safer, there are a number of successive steps that can be taken by the states concerned (either directly or in association with IMO):

- Provide proper means of establishing the position of vessels by means of lighthouses, buoys or electronic means;
- Provide pilotage (mandatory, if possible);
- Establish a TSS (Traffic Separation Scheme);
- Establish a VTS in combination with a Reporting Scheme; and
- Have vessels comply with all relevant IMO conventions and regulations, during all the previous phases.

In the area of concern, the Black Sea, most of these steps have already been taken to a certain extent. Means of establishing the position of vessels are in place in all areas leading to ports and straits. GPS (here used as a generic term for GPS/GLONASS) is available in the whole area, and in some areas Differential GPS has been established in order to enhance accuracy (e.g., Turkish Straits).

Pilotage is available for all major ports. In the case of the Turkish Straits, pilotage is available for vessels using the Strait, but this cannot, generally speaking, be made compulsory for non-Turkish vessels. Major oil companies, however, want the masters of the vessels they charter to use pilots in the Turkish Straits. By doing so they are examples for all other shipping traffic.

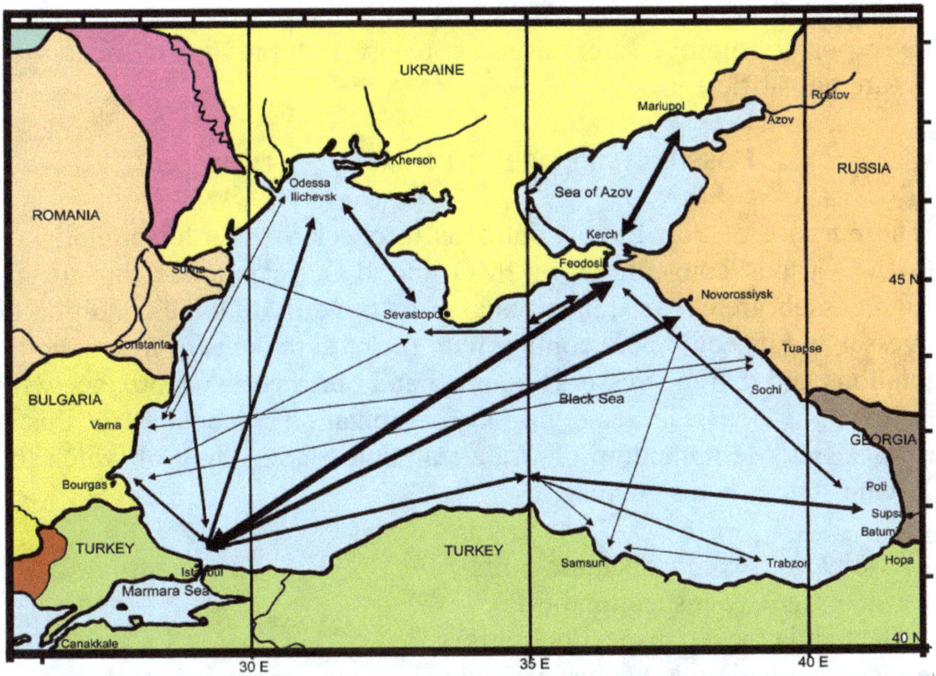

Figure 3: Most important routes in the Black Sea

According to *The Shipping Statistics Yearbook of 2003,*[6] the tanker traffic (vessels of 10,000 dead weight tons (dwt) and over) in the Black Sea area rose from 96.6 million dwt in 1999 to 113.3 million dwt in 2002, an average growth of 5.5 percent. The total number of vessels that transported this amount of cargo could not be established from the available data. Dry bulk carrier's traffic (vessels of 10,000 dwt and over) rose in the same period to 90.6 million dwt, an increase of 8.9 percent. Conventional general cargo carriers dropped to 22.6 million dwt, a decrease of 2.7 percent over the period.

The establishment of TSS has been proven to influence the traffic behaviour in a positive way and to a great extent; however, permanent observation of the vessels' compliance is necessary, which can be carried out by a VTS if the TSS is within the range of the VTS sensors. As most TSSs are under the radar coverage of a VTS, any infringements or non-compliance with rule 10 of COLREG can therefore be reported to flag States for follow-up.

Setting up and operating a VTS is the ultimate source for any coastal State or port to improve safety as a last resort if all previously mentioned items are not sufficient.

POSSIBILITIES OF PORT STATE CONTROL

The purpose of Port State Control is to check vessels for compliance with the various conventions of IMO and ILO. In practice this means foreign vessels are checked in national ports to verify that the condition of the vessel and its equipment comply with the requirements of international regulations and that the vessel is manned and operated in accordance with these rules. In the Black Sea region the organization responsible for this is the Black Sea Memorandum of Understanding[7] (BS MOU), of which the following states are a signatory:

- Bulgaria (seven inspection ports);
- Georgia (two inspection ports);
- Romania (seven inspection ports);
- Russian Federation (four inspection ports under the BS MOU);
- Turkey (five inspection ports under the BS MOU); and
- Ukraine (20 inspection ports).

The BS MOU works closely with other PSCs, such as the Mediterranean Sea MOU and the Paris MOU. Training for Port State Control Inspectors was provided by the IMO.[8] Port State Control started in the Black Sea area quite recently.[9] During 2002, 2,967 inspections were carried out and 7,020 deficiencies were reported. The number of detentions in 2002 was 168.[10]

The only available annual report of the BS MOU does not give sufficient data to compare which flag States have an excess factor with regard to the total average of detentions. If the Paris MOU annual report of 2002[11] is used, we learn that during the year 2002 the situation of flag States bordering the Black Sea was as follows for inspections at Paris MOU ports: Romania was blacklisted with an excess factor of 13.84 percent above the average of 7.98 percent; Georgia was blacklisted with an excess factor of 10.94 percent; Turkey was blacklisted with an excess factor of 10.80; Ukraine was blacklisted with an excess factor of 3.50 percent; Bulgaria was blacklisted with an excess factor of 1.59 percent;

and Russian Federation was greylisted with an excess factor of .67 percent.

If we compare the number of inspections of the BS MOU and the Paris MOU during 2002 we see the following:

MOU	Number of ships inspected in 2002	Number of deficiencies observed in 2002	Average deficiencies per vessel in 2002
BS MOU	2,967	7,020	2.36
Paris MOU	19,766	69,079	3.49

Table 1: Comparison of BS MOU and Paris MOU inspections and deficiencies found in 2002

The figures in Table 1 can only be judged properly if the number of PSC officers is also available. However, this is not the case. One can imagine that an administration with a limited number of PSC officers instructs them to concentrate on the more serious deficiencies.
If we look at the number of detentions we see the following:

MOU	Number of ships inspected in 2002	Number of detentions in 2002	Detention as a percentage of total inspected vessels in 2002
BS MOU	2,967	168 /188[12]	5.66%/ 6.33%
Paris MOU	19,766	1,577	7.79%

Table 2: Comparison of the BS MOU and Paris MOU detention rates in 2002

The Paris MOU is using selective targeting and new banning provisions, thus moving toward a "zero-tolerance" strategy.[13] In 1993 the detention percentage of the Paris MOU was 5.16 percent, which is more or less equal to the present percentage of the BS MOU. In 1993 the average number of deficiencies per vessel was 2.49, according to the Paris MOU, which again equals the present situation in the Black Sea MOU.

Port State Control is carried out in the following ports, according to the 2002 annual report of the BS MOU: State	PSC-Ports	PSC-Ports
BULGARIA	Varna Baltchik Bourgas Nessebar	Pomorie Sozopol Tzarevo
GEORGIA	Batumi	Poti.
ROMANIA	Constanta Mangalia Midia Galati	Braila Tulcea Sulina
RUSSIAN FEDERATION	Gelenzhik Novorossıysk	Sochi Tuapse
TURKEY	Kdz. Ereğli Samsun Trabzon	Rize Hopa
UKRAINE	Bilgorod-Dnistrovskyy Berdiansk Dnipro-Buzkyy Illichivsk Oyabrsk Reni Sevastopol Skadovsk Theodosia	Ust-Dunaisk Yalta Yevpatoria Yuzhnyy Izmail Kerch Kherson Mariupol Mikolayiv Odessa

Table 3: Ports with PSC

An example of a PSC report of the BS MOU on the Internet [14] can be seen in Annex 1. This particular vessel was detained by PSC, and it is interesting that the vessel transited the Turkish Straits on many occasions during the period preceding the detention.

Port State Control deserves the cooperation of all parties interested in the safety of shipping. Within the framework of this paper we can therefore mention the following entities:

- Flag States;
- Port States;
- Classification Societies;
- VTSs; and
- Pilotage.

The first three entities are quite logical. We will concentrate on the last two: VTSs and Pilotage. Pilots, who are in general the first persons to board a vessel and the last persons to disembark, often notice major deficiencies with regard to the operation of the vessel concerned and the capabilities of the crew. There should be a modus for them to report deficiencies directly to the PSC in the next port of call. However, there is one restriction: let this only be done by pilots who fall under a regime of mandatory pilotage. If there is no mandatory pilotage, pilots who report vessels not in compliance with international standards will soon notice that their voluntary services are no longer needed.

A VTS can primarily detect deficiencies with regard to the Standard Marine Phrases in a straightforward way. A VTS where master mariners are the operators, as is the case in some VTSs in the Black Sea area, can detect more deficiencies that cannot always be substantiated properly.

A VTS, if properly equipped, can also detect whether or not the AIS (Universal Shipborne Automatic Identification System) equipment is properly set up with regard to static and dynamic data. In the authors' opinion there is a case for VTS operators to inform competent authorities about these matters.

A central database restricted to those who "need to know" and used by VTS operators, PSC and other parties involved in the area could assist in this field.

An interesting outcome of the EMBARC[15] project of the European Commission is that a Risk Index has been calculated for vessels of a certain type and size-class by means of using statistical data to determine the average risk of a vessel in terms of frequency and consequences. The main consequences are pollution and loss of life. For a specific vessel, multiplication factors are determined based upon the following: age of the vessel, classification society, flag of vessel and hydro-meteorological circumstances. As a first approximation, all serious accidents and all tanker accidents during 10 years were investigated using Lloyd's database to check the ratio between percentage of accidents and the percentage of the total traffic of specific Classification Societies. The impression that well-known western European Classification Societies would have a high level of safety among their fleet was not true. The best ratio was that of the Russian Register of Shipping (0.31), the worst was for Bureau Veritas (1.92), immediately followed by Germanische Lloyd (1.81).

The age of vessels related to the casualty rate was also investigated and mapped out in some interesting graphs. An interesting conclusion was that a new vessel sailing in wind force 10 Beaufort has a larger accident probability than a 30-year-old vessel in good weather.

Another finding from the Paris MOU annual reports was that the number of detentions cannot be seen as an estimator of the probability of accident proneness. This is actually a contradiction to the fact that the results of a PSC are often measured in number of vessels detained and inspected as well as the number of deficiencies detected, as done before in a comparison between the BSMOU and the Paris MOU.

Vessel Traffic Services

Vessel Traffic Services are an important instrument for safeguarding the area where it is established. VTS is governed by guidelines of IMO with regard to the implementation and operation.[16]

A VTS performs three basic tasks in order to carry out its services:

- data collection;
- data evaluation; and
- data dissemination.

A VTS should always have a comprehensive "traffic image," which means that all factors influencing the traffic, as well as information about all participating vessels and their intentions, should be readily available. The traffic image allows for situations that are developing to be evaluated and responded to. To a great extent, the data evaluation depends on the quality of the data that is collected and the ability of the operator to combine this with an actual or developing situation. The data dissemination process consists of conveying the conclusions of the operator. Three forms of data dissemination are possible:

- Information service: providing information to vessels that, in principle, can be checked by the vessels involved; this means that at this stage the operator is not providing advice. However, in the day-to-day practice of a VTS, information that is properly given is often suggesting a specific course of action by the sequence of the information provided;
- Navigational Assistance Service: a service provided to an individual vessel at its request or when deemed necessary by the VTS to assist directly the decision-making process on board the vessel. This service consists of information relating to a specific vessel and may include warnings and advice, as long as it is not the intent to direct "the course to be steered"[17] or "engine orders to be executed".[18] In providing navigation assistance, the VTS participates in the onboard decision-making process by giving information on a vessel's course and speed made good, information on its position relative to fairway axis or waypoints, information on the vessel's position or identity and intentions of the surrounding traffic. This service is not intended to advise "courses to be steered" or "engine manoeuvres to be executed"; and
- Traffic Organization is a service to prevent the development of dangerous situations and to provide for the safe and efficient movement of traffic within the VTS area. It concerns the forward planning of movements and management of space and is particularly relevant in times of congestion or when the movement of special transports may affect the flow of the traffic. Monitoring traffic, enforcing adherence to governing rules and regulations, and communicating existing conditions to VTS participants and sometimes allied services are essential elements of traffic organization. Traffic organization may include establishing and operating a system of traffic clearances in relation to the priority of movements, allocating space, mandatory reporting of movements, establishing routes to be followed, speed limits to be observed or other appropriate measures considered necessary by the VTS authority. Where the VTS is authorized to issue traffic organization-related instructions to vessels, the instructions should be results oriented. This means that the details of execution are left to the vessel. The successful operation of a traffic organization regime

requires transparent rules, efficient communication and positive identification of all vessels.

A very important item is the acceptance of a VTS by the shipmaster and officers. A VTS operator should never consider himself to be conducting or piloting the vessel; that is the task of the master. However, if his education is at least at the same level of the master, a situation of mutual trust is easily established.

The last VTS that became operational in the Black Sea area was the Turkish Straits VTS covering Istanbul Strait and Canakkale Strait. At a later stage the Marmara Sea area will also be covered by VTS. The operators and supervisors of this VTS are all master mariners. This was a choice of the Turkish Government that shortened the training period considerably and increased automatically the mutual trust between the VTS and its customers. The training of operators and supervisors was carried out by Istanbul Technical University, Maritime Faculty at Tuzla, and carried out according to all international regulations of IMO and recommendations of IALA. The authors of this paper have been involved in this training to a large extent.[19]

Presently the Turkish Straits VTS[20] is still in a period of building up experience, using the same Traffic Management Regulations that were used in the old system. However, the operators and supervisors have already been trained to use a new system of traffic management, one that in time will become operational when the Turkish Government so decides. Therefore, the generous offer of the Russian delegation at the MSC[21] is understood as a matter of expressing good will and international cooperation.

There is a point in establishing a form of cooperation in the training of VTS operators and analyzing the various operational procedures. The Turkish Straits VTS operators are certainly willing to meet their counterparts in the other Black Sea area states to exchange knowledge and gain a better understanding of their mutual problems and associated solutions. Suggestions have already been made to organize regional VTS conferences, and this will probably take place again within the next two years. The first Black Sea Conference on training and qualifications for maritime pilots and VTS operators took place in Nikolayev (Delta-Pilot) in May 2003.

With regard to the training side of the VTS operators' profession, the following can be said. In Turkey the training has been carried out by ITU-

MF, after a "train the trainers" program conducted in The Netherlands and assisted by experts from other countries whenever necessary. There is much expertise in the field of making scenarios for the ITU-MF VTS-simulator and the execution thereof, as well as briefing and debriefing sessions. The knowledge and expertise of all regional institutes should be shared in order to enhance the level of training that can be provided.

During a regional VTS symposium the following items could provide for an interesting exchange of knowledge:

- Developments in VTS hardware and software;
- Procedures and the possibilities of harmonization;
- Exchanges of operational experience and problems encountered;
- Traffic management procedures;
- Harmonized training of operators;
- Training institutes, use of VTS simulators and VTS research; and
- Possibilities of "international twinning" as a training tool. This is already done in pilotage (working together with your counterpart, observing and learning without any formal responsibility).

The sooner a regional exchange of ideas and opinions within the field of VTS starts the better. There will always remain political differences related to traffic management in Turkish Straits and interpretations of conventions, but these difficulties should never be put upon the shoulders of the operators who have to carry out their important work.

The European Commission issued the White Paper "European Transport Policy for 2010: Time to Decide."[22] The following is from part 4:

> Lastly, the Community should gradually establish a management system for shipping off its coasts. At present, ships' movements are regulated by bilateral agreements concluded in the framework of the IMO, for instance, for the English Channel or the 'Ushant traffic separation scheme.

These local controls focus on traffic issues such as spacing, speed and routes. If the proposals already tabled by the commission in the Erika II package are adopted, they should also concern the dangerous nature of cargoes and permit the re-routing of ships in stormy weather, including those ships sailing outside of territorial waters. Irrespective of the nature of the controls, however, the information collected is generally neither used nor transmitted to the other centers, authorities or bodies along the ship route.

The (then) future European Maritime Safety Agency will facilitate systematic exchanges of information. Identification systems (transponders), black boxes on board and the Galileo programme will soon make it possible to establish a ship's position within a few meters. By 2010, the enlarged Union could thus, as in the air transport sector, have a traffic management system in place to protect itself against dangerous or suspicious movements of ships, in particular by diverting them to ports of refuge. A harmonised system of this nature for the management of shipping from the Bosphorus to the Baltic, taking in the Bay of Biscay and the English Channel, will give the European Union the means to coordinate intervention and control and thereby take effective action on the U.S. model against all hazards on its seaboard (particularly drug trafficking, illegal immigration and the transport of dangerous goods) without going so far as setting up a common coastguard.

Radar coverage of all European shores is extending. Recently a German company received an order for a coastal surveillance system to protect the Polish coast. Around 85 percent of the total cost of five million euros is financed through the E.U., which is already financing similar projects in other prospective member countries such as Bulgaria and Romania.[23]

Vessel Traffic Management and Information Service

The term Vessel Traffic Management and Information Service (VTMIS) is often used in brochures selling VTS equipment as the ultimate VTS. Some VTSs also call their services VTMIS, but this is not true according to accepted definitions. VTMIS is often seen as an extension of a VTS in the field of exchanging data and information with third parties such as ship operators, agents, ports, terminals, stevedores, bunker suppliers ship-chandlers, etc.[24]

The following is a definition of VTMIS that is used in various E.U. research projects: "VTMIS: a concept for harmonised information services to support waterborne traffic management and transport management. It includes the interfaces to other transport modes."

This means that VTMIS requires ICT-based information exchange with parties outside the VTS. A VTMIS without involvement of a VTS is possible in principle: an exchange of data between a number of ports and, for example, the coastguard.

The trend is that VTSs move into the direction of VTMIS by extending their information flows. Future members of the European Union especially have to follow all present research projects carefully in order to be able to implement harmonized procedures of VTMIS, if necessary.

Automatic Identification Systems

Automatic Identification Systems (AIS) are the ship's equivalent of secondary radar in airplanes. An AIS works on two dedicated VHF channels[25] and transmits vessels' position, course and speed over the ground to all other vessels that are similarly equipped. The position of the vessel is derived from GPS or DGPS. Depending on the status of the vessel (moored, at anchor or underway with various speeds) the data is transmitted with specific intervals.[26] Short messages can be transmitted and received as well. The range of AIS depends on the same parameters that govern VHF communication and is, therefore, under normal circumstances between 20 and 40 miles in the approaches to ports, depending on antenna height. The information received by shore-based radars is very much dependent upon the pulse length of the transmitted signal and the blocking of signals by geographical features and man-made obstructions. This is not the case with AIS. The purpose of AIS is to provide other similarly equipped vessels in the vicinity and authorities ashore with positive identification of a vessel, including name, call sign and position, and the course and speed over the ground. The aspect of a vessel (important when COLREG is involved) is not always transmitted. Many VTS centers are equipped with AIS base and repeater stations and can present the radar-image of a vessel (if available) and the AIS information on the same screen. In a VTS, AIS is mainly used when identifying a vessel approaching the VTS limits and in cases of target-swapping for re-identification. In principle, the use of AIS should decrease the amount of VHF communication taking place. The objectives of most maritime administrations are to balance safety of its waterways and protection of the environment with the efficiency of vessel traffic. AIS is an ideal tool to enforce international and national regulations, such as the proper use of TSS, and reporting schemes at distances that are often larger than the radar range of shore-based radar sensors.

However, in combination with a long-range communication medium, AIS also provides an excellent tool to meet the long-range ship tracking and monitoring requirements of a VTS. The responsibility of administrations for wide area or offshore monitoring of ship traffic includes safety of navigation, search and rescue (SAR), resource exploration and exploitation and environmental protection in offshore areas, including the continental shelf and economic exclusion zones (EEZ).[27] In certain areas tankers may only proceed in strict conformance with established Tanker Exclusion Zone (TEZ) regulations. Examples of these TEZ are as follows:

- TEZ on the West Coast of Canada; and
- a mandatory route for large tankers from North Hinder to the German Bight, and vice versa.[28]

There are two reporting systems in Australia: AUSREP and REEFREP, both adopted by IMO, which intend to use the long-range application. Adherence to these regulations must be monitored.

The national security of states with regard to terrorist threats by means of sea-going vessels is something new. AIS cannot solve this problem but can be an extra tool to determine which vessels are approaching a port and check this information with pre-arrival notices from vessels and agents available in the databases.

AIS is a new tool for vessels and VTS operators. Much experience has to be gained as more and more vessels will have this equipment. Most vessels will be equipped with AIS by the end of 2004.[29]

Presently AIS data is presented as a trial on the Internet[30] with a refreshment rate of two minutes of the entire Netherlands coast, Belgian coast and English Channel.

Also the ports of Amsterdam, Rotterdam/Europort, Flushing and Antwerp are included. Occasionally data is transmitted from Aruba and St. John, Newfoundland.

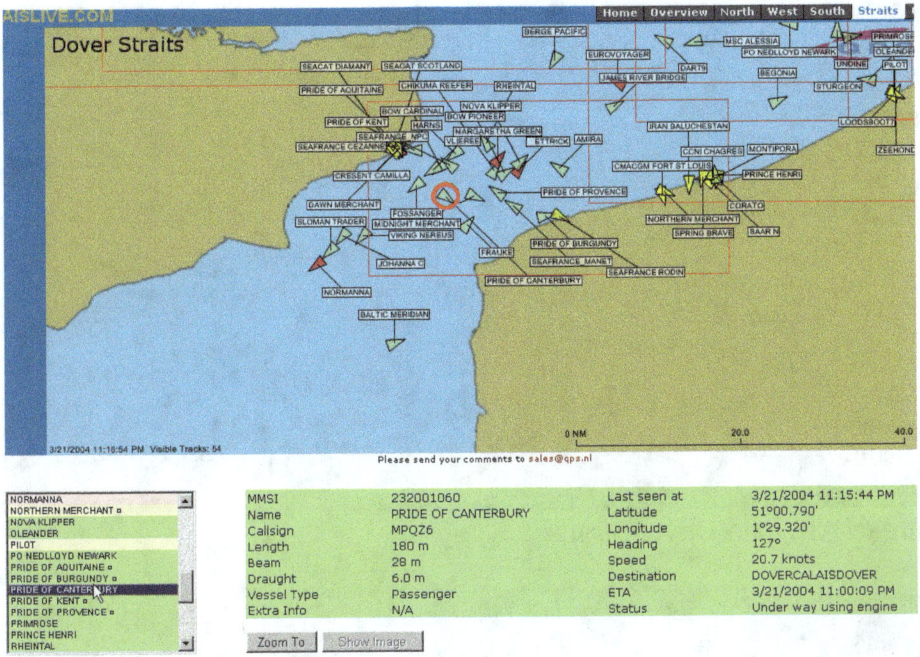

Figure 4: AIS data Dover Straits

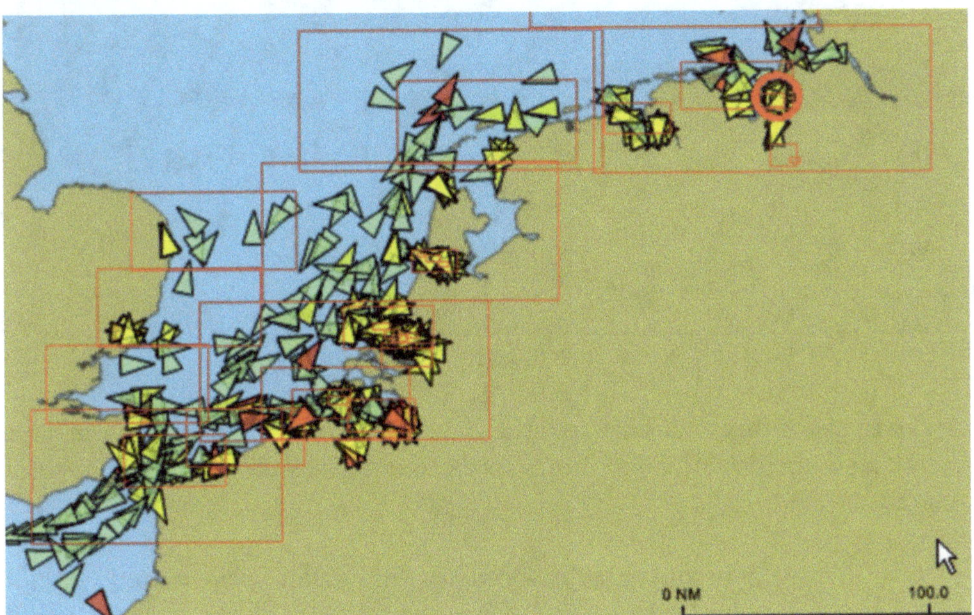

Figure 5: All available AIS data Southern North Sea

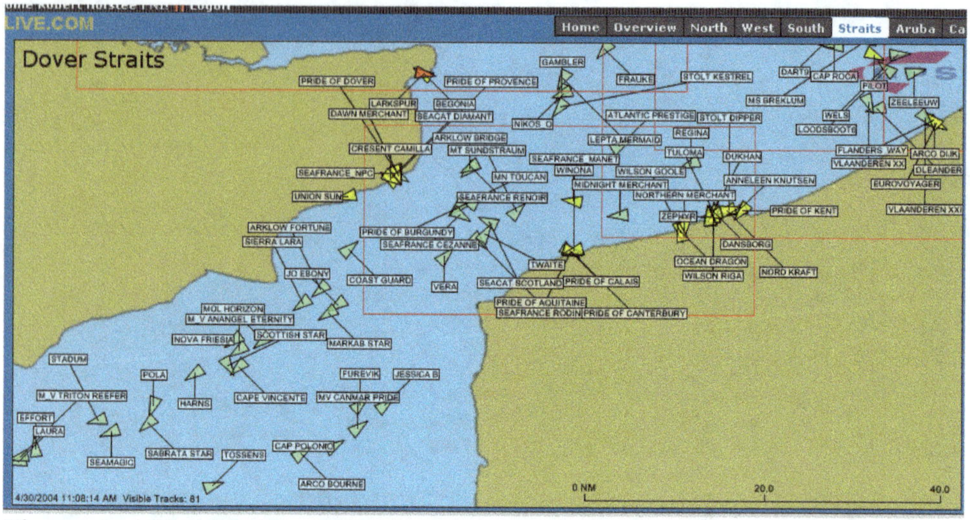

Figure 6: AIS data Dover Straits

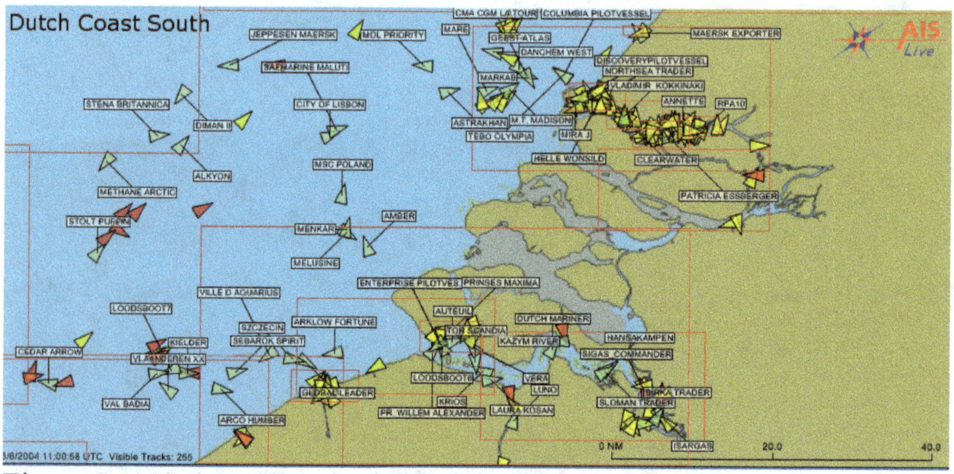

Figure 7: AIS data of Netherlands coast South

An analysis of this data on a specific date (n=217) at a time when the data could easily be downloaded into a spreadsheet gives the following general results:

- Indicator dynamic data "moored" or "at anchor" transmitted when the vessel was "under way": 4 percent.
- Indicator dynamic data "under way" transmitted when the vessel was "moored": 8 percent.

- Input of static data wrong: 5 percent (length and/or beam, 0 meters).
- Input of dynamic data draft: 2 percent unreliable or unbelievable (draft = 0 meters).
- Input of dynamic data ETA and destination wrong or doubtful: 18 percent (this includes ETA and destinations of vessels moored).
- Input of dynamic data in the field "extra info" sometimes doubtful with regard to restrictions because of deep-draft.
- Input of dynamic data with regard to hazardous cargo (categories A, B, C and D)[31] was often used, but cannot be checked on reliability.

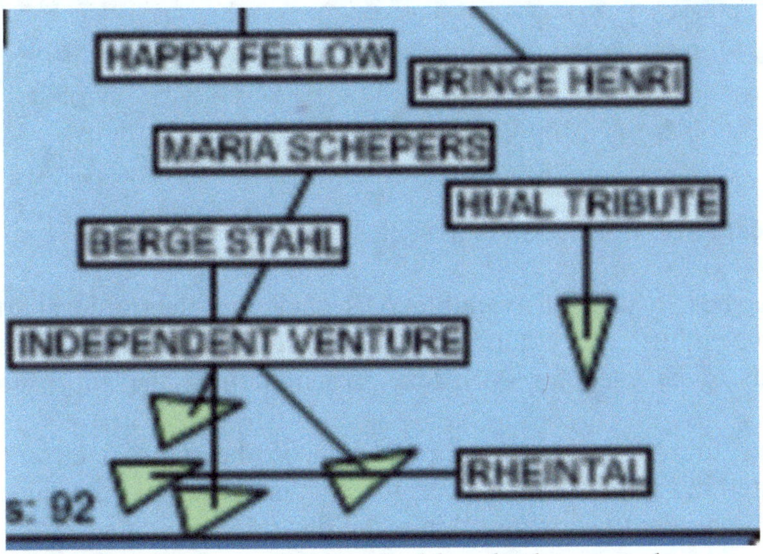

Figure 8: Close-up of Berge Stahl and other vessels

MMSI	258817000	Last seen at	6/6/2004 1:06:40 PM
Name	BERGE STAHL	Latitude	N 50°28.270'
Callsign	LATO2	Longitude	E 0°50.570'
IMO number	8420804	Heading	76°
Length	342 m	Speed	14.0 knots
Beam	64 m	Destination	ROTTERDAM
Draught	22.5 m	ETA	6/6/2004 11:00:00 PM
Vessel Type	Cargo	Status	Under way using engine
Extra Info	N/A		

Figure 9: Data received from Ore carrier Berge Stahl

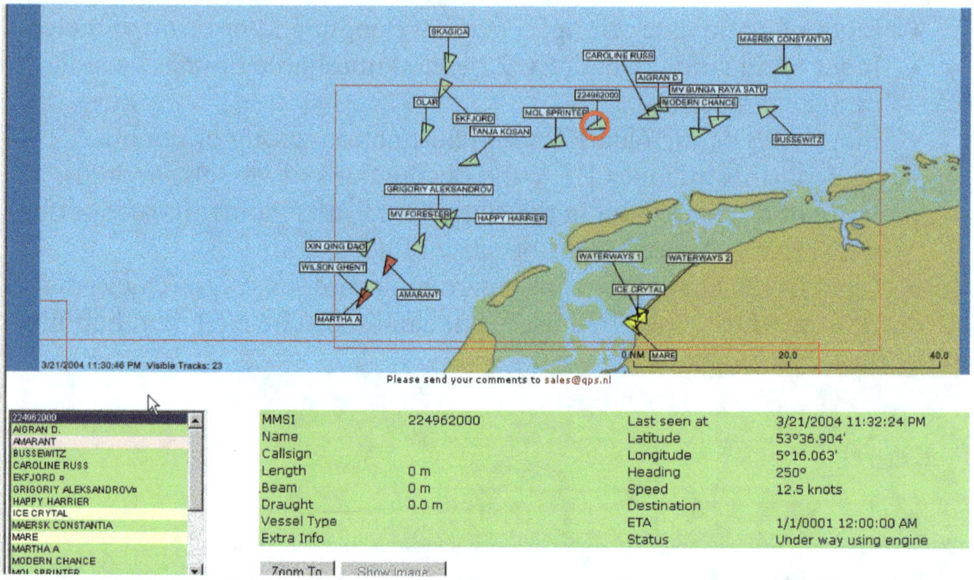

Figure 10: Critical AIS information is missing

If a VTS wishes to use all available AIS data, the electronic chart displays have to be adjusted to extend beyond the radar-coverage area or a special chart has to be used for AIS data presentation in the approach areas to the VTS.

The VTS and the AIS data can be presented on the same electronic chart, as illustrated in Figure 11. However, the data is not exactly the same by definition. The prudent user should, in principle, compare data derived from the same source.

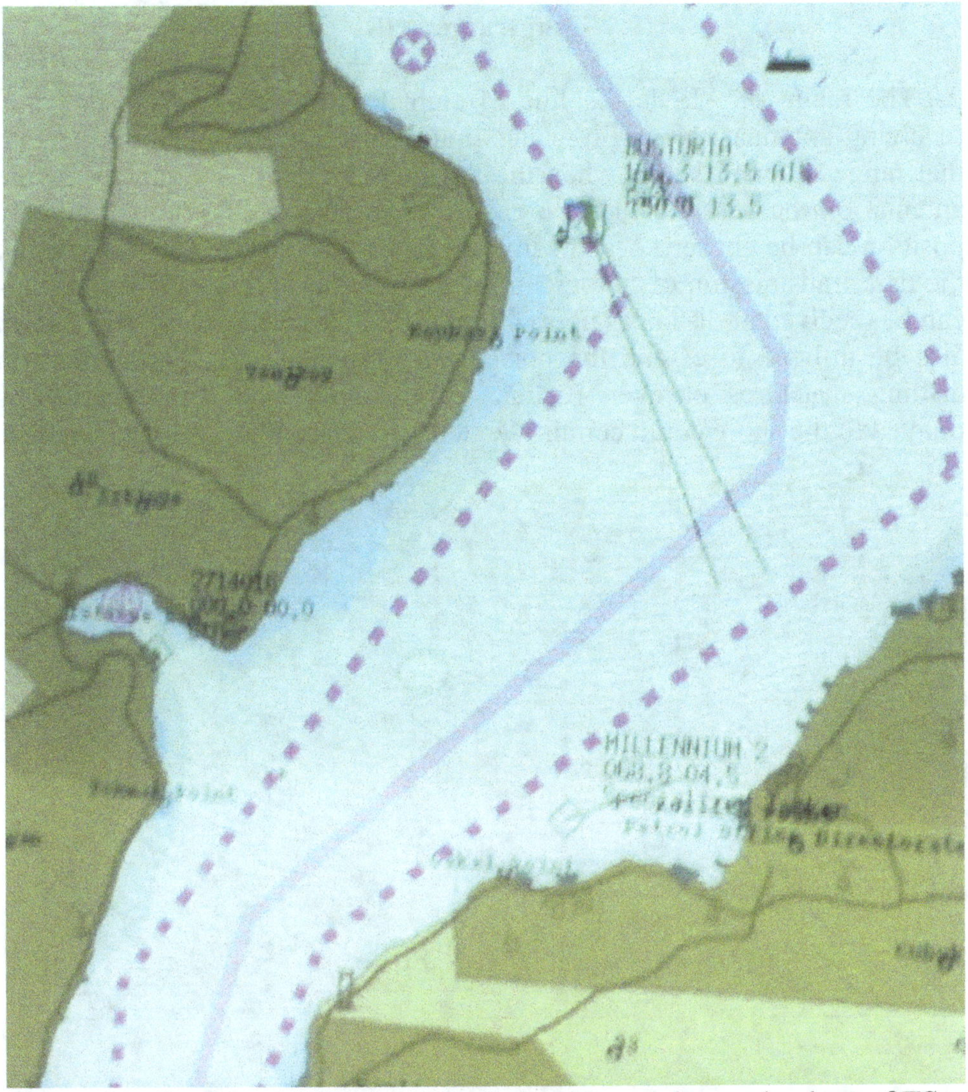

Figure 11: Combined VTS & AIS information on electronic chart of TS-VTS

The availability of real-time AIS data on the Internet could be considered a matter of great concern, if this data is used for unlawful practices. Interpretation of the available data by a layman (journalist) could lead to negative publicity of shipping in general and the responsible authorities in particular.

Long-range AIS

The range of AIS is the line of sight between the transmitting and receiving antennas. Normally a maximum range of 20 miles is used, but that range is much larger when the antennas are high. The height of the antenna on the vessel cannot be changed, but ashore an optimal antenna position can be chosen. A short investigation of available data shows that the data transmission of 50 percent of all presently AIS-equipped vessels can be received at distances up to 40 nautical miles. AIS relay-stations can also be utilized to extend the normal AIS coverage. AIS equipment on drilling structures or even regular ferries where the data received is relayed to the shore could certainly extend the present range considerably.

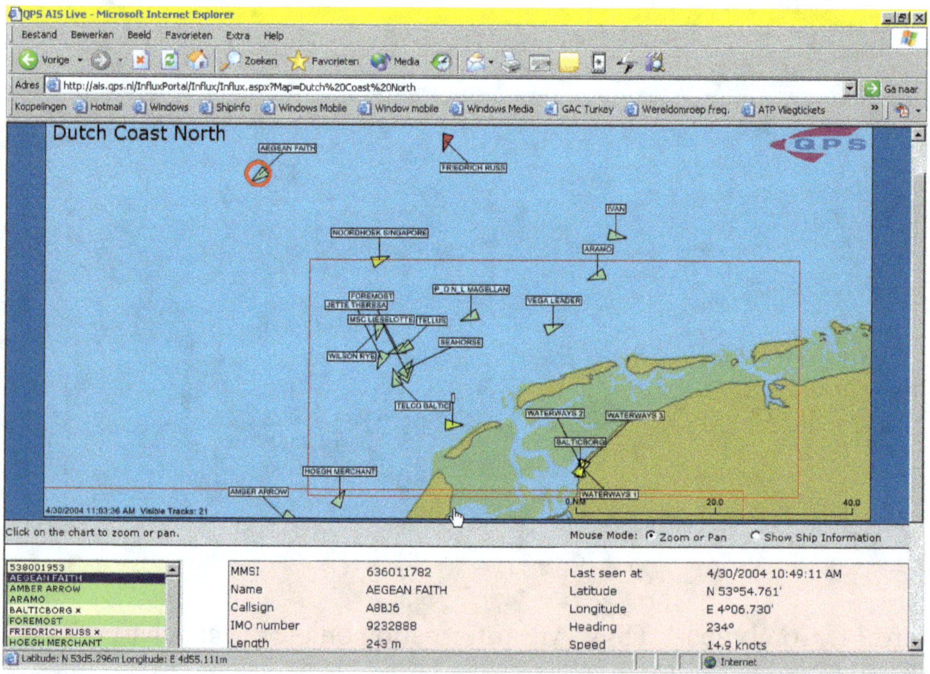

Figure 12: "Long-range coverage" near Frisian Islands during good propagation circumstances.

In the European Union several investigations have been carried out with regard to VTMIS and long-range AIS, including the Waterman Project,[32] project COMFORTABLE,[33] SEA SAFE NET and EMBARC,[34] however the results of EMBARC are not yet available to the general public. The most recent project is the MARNIS project of the 6th framework programme.

The general idea behind long-range AIS is to use satellites that interrogate the participating vessels at standard intervals. Presently, while we are in the interim period of vessels being equipped with standard AIS, not many vessels have purchased AIS equipment that has an interface for satellite communication, but this might change. The ship's security alert system, which became operational on July 1, 2004, could use the AIS/Tracs-SSAS,[35] which is also proposed for Homeland Security in the United States. The European Galileo[36] project will certainly play an important role by being an interface between vessels and the shore, related to exchange of position and security information.

VTS IN THE BLACK SEA AREA

The following VTSs exist in the Black Sea area. All ranges are the VTS-coverage ranges and not the AIS-coverage:[37]

- Port of Illichevsk, Ukraine; VTS area radius 14 nautical miles from position 46°19,1'N 30°40,8' E excluding VTS areas of Odessa and Yuzhny ports; shore-based AIS station; Consists of VTC and one remote radar station.
- Port of Odessa, Ukraine; VTS area = radius 8 nautical miles and bearing 161° from position 46°29,5'N; 30°45,6' E and meridian 30°55,0' E; shore-based AIS station; Consists of VTC and one remote radar station.
- Port of Yuzhny, Ukraine; VTS area radius 12 nautical miles from position 46°36,1' N ; 31°1,4' E. and meridian 30°55,0' E; shore-based AIS station.
- "Delta-Pilot" Regional VTS area, Ukraine; sea boundaries - meridian 31°18',7 E, parallel 46°30',0 N, estuarial waters of Bug & Dnepr to port of Nikolaev (in the northern part of area and port of Kherson in the eastern part of area); shore-based AIS station expected to be in operation by the end of May 2004. Consists of VTMC, three sub-centers and three remote radar stations.
- Sevastopol (Delta-Pilot) regional VTS area, Ukraine; parallel 44°50,4'N meridian 33°44',4E and boundaries of territorial waters (12 miles); shore-based AIS stations are planned by the end of 2004.

- Port of Kertch VTS area, Ukraine: includes Kertch-Enykal canal with southern and northern (from Azov Sea) approaches to Kertch straight and port waters; shore-based AIS station.
- Port Kavkaz (Russian Federation) VTS area: inshore fairways of Kertch Straight with southern and northern approaches; shore-based AIS station.
- Port of Novorossiisk (Russian Federation) VTS area: parallel 44°37,7'N (маяк Дообский) meridian 37°48',0 E; shore-based AIS station. Includes VTC + 4 remote radar stations. The Novorossiisk regional VTS area is being extended to Tuapse port.
- Port of Tuapse (Russian Federation) planned for commissioning in 2004; will include port approaches and anchorages.
- Supsa, Georgia (Lockheed Martin): no AIS information available.
- Constanta, Romania (Lockheed Martin): AIS information available.
- Varna, Bulgaria: no AIS information available.
- Burgas, Bulgaria: no AIS information available.
- Istanbul Strait VTS, Turkey: AIS equipped.

The present regimes oblige the master to give an ETA at least 24 to 48 hours in advance. A need for an ETA beyond this period is not needed by most VTSs.

National authorities are increasingly under public pressure to have a complete traffic image of all vessels in their area of concern, especially vessels carrying dangerous or pollutant cargo. In due time this will happen when the United States and the European Union want vessels to be equipped with satellite-enabled AIS. Even if this is the case, real traffic management cannot be executed because a real-time traffic image does not exist. Then we only know the position of a specific vessel at specific intervals (it is suggested that this interval should be in relation to the distance the vessel is positioned from the nearest coast). This information can be used for SAR purposes and for fighting terrorism.

There is a solution with using existing AIS to protect the coastal regions of states: Establish a number of AIS receivers at strategic and high locations that can relay information gathered via the Internet to the nearest national VTS center. At a later stage this information can be shared with other states. Publishing the information on a Web site, which is done in The Netherlands, is not recommended. Software can be used to notify centers that a specific boundary line has been crossed, thus generating an alarm.

It is a pity that the major routes in the Black Sea cannot all be covered 100 percent. If a small number of AIS receivers are put at strategic positions along the coast in addition to existing VTS-AIS transceivers, the entire coast of the Black Sea area could easily be covered by AIS. This data should then be transmitted via the Internet to a central position where available software should translate this data into an electronic traffic image of the coastal area. Data should be available for all parties involved in the safety of shipping, such as government agencies, PSC, pilotage, VTS, SAR and oil response teams.

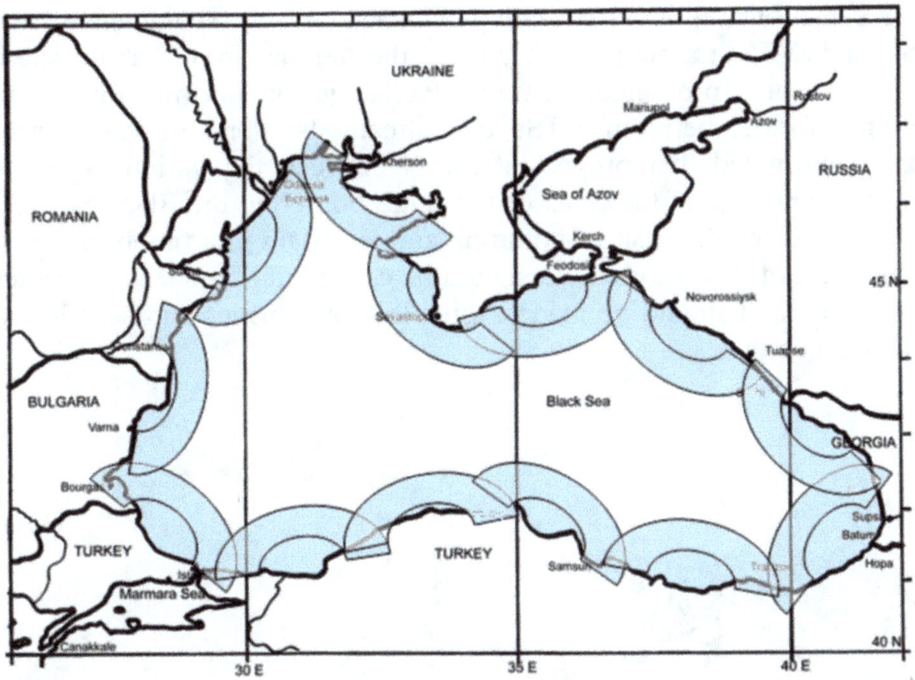

Figure 13: Possible AIS coverage of Black Sea coast

The 14 VTS centers in the Black Sea region are not yet included in the World VTS Guide.[38] The Turkish Straits VTS will be included soon. The knowledge about all VTSs in the Black Sea region should be promulgated to all concerned with safety in this region.

CONCLUSIONS AND RECOMMENDATIONS

Regulations dealing with maritime safety are becoming more and more complex. Future member States of the E.U. will have to apply the European regulations as well. It is therefore advocated that all maritime agencies of the Black Sea states follow closely any changes in regulations that apply to their own states as well as other states bordering the Black Sea. States are working closely together in Port State Control in the Black Sea in order to control and improve the safety level of all vessels plying through their waters.

Pilotage, VTS and PSC should work closely together in order to improve the overall level of safety. Cooperation is recommended for VTSs and VTS training institutes in the region in order to share experiences and promulgate safety. Exchange of information in an electronic format between VTSs can already be done in the present situation, provided that proper interfaces between the various systems exist. AIS coverage of the coast of the states bordering the Black Sea can be a tool to detect the risk of pollution and terrorism in a timely manner. Management of this information should be carried out by one VTS center in each state and should be made available to authorities and search and rescue workers.

ANNEX 1.
A report of a PSC investigation that led to a detention, taken from the Internet site of the BS MOU.

IMO #	7364637	Ship name	ERDAL KARAER
Flag	Turkey	Owner	PASIFIK DENIZCILIK SAN, VE DIS TIC. LTD. STI
Year of build	1973	GT	1110
Reporting Authority	Romania	Port of detention	Constanta
Date of det.	731490	Date of release	
Ship type	oil tanker	Ship class	Turkish Lloyd
Nature of def. (Responsible RO, if any)	CERTIFICATION AND WATCHKEEPING FOR SEAFARERS, certificates for master and officer, excepting ch.engineer, all officers without tanker specialization certificate (including master) CERTIFICATION AND WATCHKEEPING FOR SEAFARERS, certificate for ratings for watch keeping, two a.b. and wiper without tanker familiarization certificate SAFETY OF NAVIGATION, echo-sounding device, out of order FIRE FIGHTING MEASURES, personal equipment, e.e.b.d.s missing STABILITY, STRUCTURE AND RELATED EQUIPMENT, enhanced programme of inspection, enhanced survey file missing LIFESAVING APPLIANCES, line throwing appliance, one pcs. missing STABILITY, STRUCTURE AND RELATED EQUIPMENT, decks - corrosion, main deck and deck pipes corroded PROPULSION AND AUXILIARY MACHINERY, other (machinery), m.e. and d.g. high pressure fuel pipes without double skin system LOAD LINES, manholes/flush scuttles, rubber gaskets from cargo tanks manholes damaged ISM RELATED DEFICIENCIES, safety and environmental policy, safety management system not implemented		

Nota Bene: The vessel Erdal Karaer made the following transits via the Turkish Straits between 20/07/2002 and 19/09/2003, a period of 14 months: four northbound passages Canakkale Strait (of which one passage had a pilot) and 24 southbound passages Istanbul Strait (of which two passages had a pilot). Detention per 02/10/2003 not released as of 31/12/2003. Southbound passage information via Canakkale Strait and northbound passages via Istanbul Strait were not available.

ANNEX 2.
Identifiers used in combination with ship-type for cargo:
XY
X: ship type (tanker, passenger ship, etc.)
Y: hazardous cargo (category A, B, C, or D)
1– Carrying DG, HS, or MP IMO hazard or pollutant category A
2 – Carrying DG, HS, or MP IMO hazard or pollutant category B
3 – Carrying DG, HS, or MP IMO hazard or pollutant category C
4 – Carrying DG, HS, or MP IMO hazard or pollutant category D

DG = Dangerous Goods; HS = Harmful Substances; MP = Marine Pollutants

Maximum discharge quantity			
Pollution Category	Existing ships	New Ships	Conditions of discharge regulated (Yes/No)
A	None	None	Not applicable
B	300 litres	100 litres	Yes
C	900 litres	300 litres	Yes
D	Unrestricted	Unrestricted	Yes
Appendix III	Unrestricted	Unrestricted	No

Table 4 Discharge information transmitted by AIS

Notes

[1] IMO conventions:
The majority of conventions adopted under the auspices of IMO or for which the Organization is otherwise responsible fall into three main categories. The first group is concerned with maritime safety; the second with the prevention of marine pollution; and the third with liability and compensation, especially in relation to damage caused by pollution. Outside these major groupings are a number of other conventions dealing with facilitation, tonnage measurement, unlawful acts against shipping and salvage.

[2] IMO MSC 78/INF.16.

[3] SOLAS/CONF.5/34 Annex 1.

[4] EMSA: European Maritime Safety Agency <http://www.emsa.eu.int>.

[5] Commission of the European Communities, *Directive 2002/59/EC of 27/06/2002 Establishing a Community vessel traffic monitoring and information system and repealing Council Directive 93/75/EEC.*

[6] ISL, *Shipping Statistics Yearbook 2003*, p169.

[7] *Memorandum of Understanding On Port State Control in the Black Sea Region*, signed 07-04-2000 Istanbul Turkey.

[8] Reference to report of IMO re. Training of PSC officers training.

[9] BS MOU, Annual report year 2002.

[10] *Ibis.*

[11] Annual Report 2002, Paris Memorandum of Understanding.

[12] Two different total figures are given in respective tables: 168 (p 11) and 188 (p 13).

[13] OJ L019, 22.01.2002. DIRECTIVE 2001/106/EC OF THE EUROPEAN PARLIAMENT AND OF THE COUNCIL of 19 December 2001 amending Council Directive 95/21/EC concerning the enforcement, in respect of shipping using Community ports and sailing in the waters under the jurisdiction of the Member States, of international standards for ship safety, pollution prevention and shipboard living and working conditions (Port State Control).

[14] <http://www.bsmou.org>.

[15] <http://www.euro-embarc.com/>.

16 IMO A. 857(20) previously A. 578(14) which was revoked on 27/11/1997

[17] Terminology according to IMO A. 857(20)

[18] *Ibid*

[19] Hofstee et al, *Symposium Proceedings of 10th International Symposium on VTS*, Hong Kong, February 2004, session 4: Training program of Turkish Straits VTS operators.

[20] IMO, MSC 78/INF.16, dated 19 March 2004. Submission by Turkey.
Paragraph 14:

> The efficiency and success of the services provided by the TSVTS system and its positive effects on the safety of navigation in the Turkish Straits during this short period has confirmed once again the crucial importance of the VTS for the Turkish Straits. The Government of Turkey shall continue to ensure the quality of the services by updating the technological infrastructure and continued systematic training of the operational personnel and inform all users and IALA accordingly. It is believed, however, that maximizing navigational safety and maritime security and minimizing risks needs cooperation and full compliance by shipmasters during their passage through the Turkish Straits and their making use of pilotage service, as is strongly recommended by IMO resolution A.827(19).

[21] IMO, MSC 78/25/4, dated 5 March 2004. Submission by the Russian Federation.
"The Russian federation is fully confident that the situation in the Black Sea Straits can be improved with the view of effective and duly managed vessel traffic in the interest of safety of navigation and the protection of maritime environment. This must be achieved first of all by additional professional training of the VTS operators and in combination with utilizing the modern technologies (such as ECDIS and AIS) onboard ships and ashore to provide continuous vessel traffic control and better information for ships.'
'The Russian federation is open to co-operate with the Turkish Maritime Administration in this area and ready to assist in advanced training of the VTS operators at the Russian VTS Centers."

[22] European Commission, White Paper — European transport policy for 2010: time to decide, Luxembourg: Office for Official Publications of the European Communities, ISBN 92-894-0341-1.

[23] *Digital Ship* May 2004 p6.

[24] Koopmans, *Symposium Proceedings of 10th International Symposium on VTS*, Hongkong, 10-13 February 2004, session 5, paper 1.

[25] The channels are AIS 1 - 87B (161.975 MHz) and AIS 2 - 88B (162.025 MHz).

[26] AIS shall:

- provide automatically to appropriately equipped shore stations, other ships and aircraft information, including the ship's identity, type, position, source, speed, navigational status and other safety-related information;
- receive automatically such information from similarly fitted ships;
- monitor and track ships;
- exchange data with shore-based facilities;
- the requirements (AIS always to be switched on...) shall not be applied to cases where international agreements, rules or standards provide for the protection of navigational information; and
- AIS shall be operated taking into account the guidelines adopted by the Organization. Ships fitted with AIS shall maintain AIS in operation at all time except where international agreements, rules or standards provide for the protection of navigational information.

[27] *IALA Guidelines on Universal Shipborne Automatic Identification Systems.*

[28] IMO document MSC 67/22/Add 1 - Annex 11.

[29] IMO AIS Implementation dates as of 20/12/2002:
All ships of 300 gross tonnage and upwards engaged on international voyages and cargo ships of 500 gross tonnage and upwards not engaged on international voyages and passenger ships irrespective of size shall be fitted with an automatic identification system (AIS), as follows:

- ships constructed on or after 1 July 2002;
- ships engaged on international voyages constructed before 1 July 2002:
    - in the case of passenger ships, not later than 1 July 2003;
    - in the case of tankers, not later than the first survey for safety equipment on or after 1 July 2003;
    - in the case of ships, other than passenger ships and tankers, of 50,000 gross tonnage and upwards, not later than 1 July 2004;
    - in the case of ships, other than passenger ships and tankers, of 300 gross tonnage and upwards, but less than 50,000 gross tonnage, not later than the first survey for safety equipment after 1 July 2004, or by 31 December 2004, whichever occurs earlier; and
    - ships not engaged on international voyages constructed before 1 July 2002, not later than 1 July 2008;

The Administration may exempt ships from the application of the requirements of this paragraph when such ships will be taken permanently out of service within two years after the implementation date specified in subparagraphs .2 and .3;

[30] <http://ais.qps.nl>.

[31] see Annex 2.

[32] <http://www.waterman-ts.net>.

[33] <http://www.cordis.lu/transport/src/comfort.htm>.

[34] <http://www.euro-embarc.com>.

[35] Maritime Journal, March 2004, p.50.

[36] MSC 78/11.5 dated 27.02.2004 submission by the European Commission.

[37] Input for Ukraine and Russian Federation from Capt. Valeriy Latypov (Delta Pilots).

[38] <http://www.worldvtsguide.org/>.

Oil Spill Contingency Planning and Technical Cooperation of the Black Sea Region

John Østergaard[*]

INTRODUCTION

For almost 10 years the International Maritime Organization (IMO) has supported the development of modern national contingency plans for the Black Sea Coastal States. In addition to this work IMO has been responsible for the development of the Regional Black Sea Oil Spill Contingency Plan. As a result of this technical support, Bulgaria, Romania and Russian Federation have now developed, and in most cases implemented, new national plans. Georgia, Turkey and Ukraine are in the process of developing or finalizing their national plans.

REGIONAL COOPERATION

In 1994 IMO was invited to take part in the implementation of the Black Sea Environment Programme (BSEP), then established as a five-year programme, funded by the Global Environment Facility (GEF) and supported by UNDP, UNEP and the E.U. Tacis Programme.

The Ministerial Meeting of Ministers of the Black Sea States responsible for protecting the environment of the Black Sea adopted the Ministerial Declaration on the Protection of the Black Sea – The Odessa Declaration - on April 7, 1993, in Odessa, Ukraine.

The Ministerial Declaration called for national plans for applying MARPOL Special Area requirements, the enhancement of port reception facilities for ship-generated wastes and the development of national and regional contingency plans as identified by the Convention on the Protection of the Black Sea against Pollution, 1992 – the Bucharest Convention, 1992.

---

[*] Senior Adviser on Marine Pollution, Marine Environment Division, International Maritime Organization.

During 1995, IMO undertook a fact-finding mission to all major ports of the Black Sea to investigate the current status of port reception facilities and national contingency planning. At its meetings in 1995 and 1996, the Emergency Response Expert Group established under BSEP considered the findings of the IMO mission and agreed that regional cooperation on oil spill response had to be built on modern implemented national oil spill contingency plans developed in accordance with the IMO Guidelines on Oil Spill Contingency Planning.

At a Ministerial Conference held in Istanbul, Turkey, in October 1996, the six coastal States adopted the Strategic Action Plan for the Rehabilitation and Protection of the Black Sea.

The Action Plan considers vessel source pollution, including Special Area requirements under MARPOL 73/78; the establishment of port reception facilities in all ports, a harmonized system of port State control; and enforcement, including fines for illegal discharges and the avoidance of further introduction of exotic species.

With regard to contingency planning and emergency response, the Black Sea Commission was recommended to develop a strategy before the end of 1997 that would ensure that national and local contingency plans were developed before the end of 1998 and also to ensure that a Regional Contingency Plan was developed and adopted before the end of 2000.

In order to implement the Action Plan, the Black Sea Commission was established as a succession for the GEF/UNDP/UNEP/E.U.-funded BSEP Secretariat that completed its work during 1997. Unfortunately, the Black Sea States did not pay their agreed contributions to the Black Sea Commission, and only through special grants from UNDP and some donor countries did the Commission manage to "survive" until 1999, when additional grants were provided by external donors with the condition that the Black Sea States paid their contributions as originally agreed in 1996. As a consequence of the lack of funding, activities related to regional and national oil spill contingency planning were also postponed.

In early 1999, IMO was contacted by the Black Sea Commission and the international oil industry and again invited to facilitate the development of a regional oil spill contingency plan and, as appropriate, the development of national oil spill contingency plans. The IMO TC Fund provided the necessary funding, and during the autumn of 1999 and spring of 2000, two Regional Workshops on the development of a Regional Plan were held in Varna, Bulgaria, and in Constantza, Romania, respectively. The second workshop approved, from a technical point of view, the draft Regional Contingency Plan together with an associated draft legal document to the Protocol on Cooperation in Combating Pollution of the Black Sea Marine Environment by Oil or Other Harmful Substances in Emergency Situations to the Bucharest Convention.

At its meeting in early 2001, the Black Sea Commission could not agree to the draft legal document or the draft Regional Plan and decided to return the draft documents to the Advisory Group on the Environmental Safety Aspects of Shipping (AG-ESAS) for further consideration. AG-ESAS was established as a subsidiary body of the Black Sea Commission. Again IMO was invited to facilitate the work, and at an April 2002 meeting the draft Plan was finalized and approved with a view to resubmitting it to the Black Sea Commission for adoption.

The Commission considered and approved the revised draft Plan at its meeting in July 2003. By the end of that meeting, Bulgaria, Romania and Turkey signed the Plan. It is expected that Russian Federation will sign the Plan during 2004. With regard to Georgia, IMO has unofficially been informed that its parliament approved the signing of the Plan in late 2003. However, with the latest political developments in Georgia it has not been possible to confirm this, as a major reorganization of the Georgian Government is still in progress.

Paralleled with the development of the draft Regional Plan, IMO initiated a regional training programme in 2000 in cooperation with the International Oil Industry by means of the IMO, OPRC Model Courses, levels 2 and 3. All Black Sea states were invited to host the training courses, but only Russian Federation reacted positively and a successful cooperation with the State Maritime Academy in Novorossiysk was initiated.

From 2000 to the end of 2004, IMO has at the regional level organized four OPRC level 2 courses and four level 3 courses with an average attendance of three participants from each of the Black Sea states (18 students with an additional 10 to 12 students from Russian Federation). In total, approximately 200 students have attended the courses. Many of the students have attended both levels 2 and level 3 courses. In addition, IMO has organized three OPRC level 2 courses and four national OPRC level 3 courses at the national level. These training activities are having a positive influence on the development of national contingency planning and are contributing to a better understanding among the states at the regional level.

In order to maintain this positive development in the future, IMO intends to continue its support of the Black Sea region by organizing two annual courses, one OPRC level 2 course and one level 3 course.

In addition to the OPRC training courses, IMO, together with the U.S. Department of Energy, organized a Regional Oil Spill Modelling training course in January 2001 in Tbilisi, Georgia. Twenty-four participants representing all of the Black Sea states attended this course.

Further, IMO organized a regional training course on MARPOL Implementation and Enforcement in March and April 2003 in Batumi, Georgia. The 27 participants were from Black Sea states with the exception of Ukraine.

National Activities

BULGARIA

Bulgaria is in the process of revising its National Oil Spill Contingency Plan developed in the late 1990s. The present Contingency Plan has been developed in compliance with the IMO Guidelines. However, due to major reorganization of the government and its agencies, it has been necessary to revise the Plan in order to reflect changes in responsibilities of the National Administration.

In order to facilitate revision of the plan, IMO organized an OPRC level 3 course in September 2003 in Varna. More than 30 persons representing all relevant administrations and the national oil industry participated in the course. IMO is planning to provide an OPRC level 2 course during 2004.

Bulgaria is host to the Emergency Response Activity Centre (ERAC) under the Black Sea Commission. The objectives of ERAC are to play a regional coordinating role for all activities related to ship-related pollutions of the Black Sea. In order to facilitate capacity building at ERAC, IMO has paid special attention to that regional body.

GEORGIA

In Georgia, a Dutch consultant has developed a first draft of a National Plan under the World Bank's (WB) Georgian Coastal Zone Management (GCZM) Project, Phase I. The Terms of Reference for this work was developed by IMO in 1998 during the appraisal phase of the WB Project.

During a planned second phase of this project, the draft National Plan for Georgia was expected to be further developed and implemented. However, the Plan has not yet been finalized and is being considered by World Bank consultants in cooperation with national responsible authorities.

As the designated National Responsible Authority for oil pollution response, the Georgian Maritime Transport Administration (GMTA) asked IMO for assistance to finalize and implement the draft Plan in the autumn of 2001 in order to continue the national process. In November 2001, the draft National Plan was submitted to IMO for review. IMO submitted comments and suggestions for the Plan's improvement to GMTA in late January 2002. IMO was further invited to consider additional assistance, including training of personnel and technical assistance to finalize and implement the draft Plan.

In response to the request, IMO organized National OPRC level 3 and 2 courses in May and November 2002, respectively. Approximately 30 participants from relevant administrations and the oil industry participated in both courses.

In addition to the training, on two occasions IMO provided technical advice on how to proceed in finalizing the National Plan. The first advice was provided in May 2002. The outcome of this meeting resulted in agreement on substantial amendments to the draft National Plan. After further national consultations, the amended Plan was submitted to IMO in December 2002 for review and comments. The IMO review resulted in a complete reorganization of the content of the draft Plan.

The outcome of the IMO review was discussed during a two-day meeting held in April 2003 at the Ministry of Environment and Natural Resources in Tbilisi. The meeting resulted in an agreement to establish a National Oil Spill Planning Group with representatives from the key players at the national level. A major obstruction to the development and agreement to a National Plan is caused by current national legislation, which is unclear with regard to national responsibilities in contingency planning and response. Consequently, the legislation is interpreted differently by the two key ministries.

The status of the National Plan is unclear, but it is assumed that the political unrest, general election and the formation of a new government in 2004 delayed the process for several months.

ROMANIA

Romania developed a revised National Oil Spill Contingency Plan in the early 1990s. The Plan was never effectively implemented, and with the country's application for European Union membership, it agreed to review the Plan in order to ensure that it was in full compliance with the European and international requirements.

This review is ongoing. Funding has been provided under the Tasis Programme and through bilateral support from the Netherlands. A Dutch company is in the process of updating the Plan to cover both response to oil and noxious hazard substances (NHS). The Tasis Programme also includes the development of a national training capability on oil spill response and HNS.

IMO conducted a National OPRC level 3 course in May 2003 in Constantza and planned to conduct a National OPRC level 2 course during the last half of 2004.

In order to harmonize the IMO activities with E.U.-funded Tasis project and the assistance undertaken by the Netherlands, IMO, at the request of the Romanian Maritime Administration, gave a number of presentations on issues related to HNS Convention and the HNS Protocol to the OPRC 1990 at a national training course in April 2004.

RUSSIAN FEDERATION

The Ministry of General Emergency Planning of the Russian Federation, which is the National Responsible Authority for all kinds of emergency planning, has delegated the response to oil spills at sea to the Marine Pollution Control & Salvage Authority (MPCSA) under the Federal Ministry of Transport and Communications.

The MPCSA initiated the development of a modern National Oil Spill Contingency Plan (NOSCP) in the late 1990s. This new NOSCP, which is in compliance with the IMO Guidelines on Contingency Planning, was officially adopted during the autumn of 2003. Parallel with the development of the NOSCP, work was initiated to develop several regional plans for the Federation. As far as IMO is aware, modern plans have been developed, officially adopted and implemented for the following sea areas: the Baltic Sea, the Black Sea and the Caspian Sea. Work has been initiated on the development of regional plans for the Sakhalin Region, the Pacific Region and the Barents Sea Region. The development of a plan for the Siberian Arctic Sea Region is expected to be initiated within the next year or two.

With regard to the Black Sea Regional Plan of the Russian Federation, IMO has, in conjunction with the Regional Courses, provided advice and guidance. All key personnel involved in the regional planning process have attended the Black Sea Regional levels 2 and 3 courses organized by IMO in Novorossiysk. In addition to representatives of the Russian local Regional Authorities, representatives of Regional Authorities from other parts of the Russian Federation, as well as representatives of the Russian oil industry and pipeline companies have also attended the IMO organized courses in Novorossiysk.

TURKEY

Turkey has not yet developed a National Oil Spill Contingency Plan. It is assumed the main reason for this is that the Turkish Parliament has not yet adopted the necessary national legislation in this regard. According to information provided by the Turkish representatives participating in the Black Sea Regional Co-operation, it is expected that the Turkish Parliament will adopt the necessary national legislation during 2004 and will designate one National Responsible Authority with regard to oil spill contingency preparedness, response and cooperation.

It is further expected that one of the first tasks designated to this National Responsible Authority will be the development of a National Plan that harmonizes the existing contingency plans developed by a number of regional authorities.

During informal consultations with representatives of various Turkish authorities involved in regional and national oil spill contingency planning, IMO has been invited to provide assistance as soon as the required national legislation has been adopted.

Due to the controversy on oil transport through the Turkish Straits and the ongoing construction of the Baku–Tbilisi–Seyhan oil pipeline, the International Oil Industry is prepared to provide support for any IMO activities to be initiated in Turkey through its Oil Spill Preparedness Regional Initiative (Caspian Sea, Black Sea, Central Eurasia) OSPRI.

With this background IMO has tentatively allocated TC funds for one OPRC level 3 training course and one OPRC level 2 training course to be implemented during 2004 and 2005.

UKRAINE

National oil spill contingency planning in Ukraine is the responsibility of the Ministry of Emergencies and Affairs of Population Protection from the Consequences of the Chernobyl Catastrophe. Ukraine has received substantial support in funding and advice from various U.S. authorities and private contractors on the development of its National Oil Spill Contingency Plan. Unfortunately, the National Plan does not fully consider the regional cooperation developed under the umbrella of the Bucharest Convention, and a number of mechanisms necessary to facilitate regional cooperation are omitted.

This problem surfaced clearly during a regional "table top" exercise organized and initiated in November 2003 in Odessa, in which representatives of the National Responsible Authorities from all the Black Sea coastal States and IMO participated.

The main national conclusion of the Ukrainian authorities following this exercise was that the National Plan has to be reviewed and amended in order to ensure that the plan is compatible with the Black Sea Regional Plan and the national plans of the other Black Sea states.

With this conclusion in mind, IMO has received a request for assistance from the Responsible Ministry to provide technical assistance in the form of training personnel directly involved in national and local oil spill contingency planning. In response, IMO has tentatively planned for one OPRC level 3 training course and one level 2 training course to be held during 2004 and 2005. In addition to the organisation of the training courses, IMO will provide advice and assistance as appropriate in order to ensure that the necessary amendments to the National Plan are developed and implemented as soon as possible.

CONCLUSIONS

Although only very limited funding and personnel have been available, the IMO technical assistance programme for the Black Sea region has been reasonably successful and has, from an oil spill response point of view, had a major impact on the preparedness, response and cooperation of the region.

IMO, the Black Sea coastal States and the International Oil Industry have cooperated closely as equal share holders or partners in the regional contingency planning process and have all contributed to its success.

In conclusion, the success can be summarized as follows:

- All Black Sea states have in place or are in the process of finalizing a modern national oil spill contingency plan based on internationally adopted guidelines;

- A Regional Oil Spill Contingency Plan has been adopted and is expected to enter into force within the next year;

- More than 200 persons representing government agencies and the oil industry have received training in oil spill preparedness, response and cooperation;

- The IMO/Oil Industry Global Initiative has created a situation of self-sustainability in the region and the momentum that exists can be maintained by the designated authorities with a minimum of external support;

- The IMO/Oil Industry support to the Black Sea region will continue at a reduced level and will be subject to availability of resources, in particular the necessary funding; and

- A number of the functions carried out by the IMO/Oil Industry Global Initiative to date will continue to be performed by Oil Spill Preparedness Regional Initiative (Caspian Sea, Black Sea, Central Eurasia) OSPRI.

PANEL V: LEGAL ASPECTS OF ENERGY DEVELOPMENT AND DISTRIBUTION

Energy Development and Distribution – What Can the Law Do?

Paul B. Stephan[*]

Large capital investment and huge economic risk accompany energy development and distribution. Moreover, these projects often implicate multiple jurisdictions, with the location of the resource, the path of the distribution system, the source of the capital, and the destination of the product involving different states. Law faces a great challenge in dealing with these problems and the competing claims of various sovereigns to govern different aspects of the transaction.

In asking what law might do to meet this challenge, I wear several hats. As a legal theorist, I offer some reasons why law can affect the costs of energy projects. As a student of international law, I discuss some particular issues that the multijurisdictional dimensions of these transactions present. As a transactional lawyer, I identify specific problems that have arisen in the past and that counselors should anticipate will arise in the future. Finally, as a realist I counsel that law's role, although valuable, is limited and that no one should fall prey to the nirvana fallacy when considering what law can do for energy transactions.

I

Theorists accept that one of the principal functions of a legal regime is to manage the risk of opportunism in transactions requiring cooperation over time among persons with conflicting interests. In spot transactions, law's role is limited to the provision of minimal levels of security, so that transactors can complete a simultaneous exchange without fear of harassment. Complex, long-term investments, in contrast, invite opportunistic behavior, and energy projects exemplify this problem. Consider a company that absorbs the costs of prospecting for energy and, after promising studies, builds several offshore drilling platforms. Hundreds of millions of dollars may be disbursed before the first barrel of

* Lewis F. Powell, Jr. Professor of Law and Hunton & Williams Research Professor, University of Virginia.

product reaches the market. Add to this a challenging distribution issue, such as the need to build a pipeline, loading structures or gas liquification facilities. Again, huge sums of money must be converted into static, impossible-to-resale assets before the project produces any return.[1]

Imagine, for the sake of exposition, that the transaction involves two parties: an energy firm that will explore for the resource and build the infrastructure for producing and distributing the product, and the sovereign state on whose territory the resource exists. In a thought experiment, assume that these two parties come to an agreement about their mutual rights and responsibilities, but no law exists to enforce their understanding. The firm will spend, hypothetically, $1 billion to create a production and transportation structure. Further assume that the structure has value only in place and only if used as intended: No one can tear it up and sell off the parts for anything approaching the cost of creating the structure, but someone else can take it over and operate it at little additional cost. Further assume that both parties are uncertain about the future market prices of energy products. Let us then assume that the host country agrees that the firm can keep 60 percent of the oil it produces over the 15-year period beginning from the moment that production begins, with the balance going to the host country as compensation for the surrender of its rights in the resource. At the end of this period, the host country will acquire full ownership of everything associated with the project.

What could go wrong? From the perspective of the firm, it may expend great sums in creating the production and transportation structure, only to be ousted by the host country at the point where operating the project becomes profitable. From the perspective of the host country, it may have gotten too little for what it gives up, especially if future energy prices rise. The firm worries about expropriation; the host country fears the firm will hide and steal production. The rational decision is to refuse to undertake this potentially profitable project absent some way of managing the risk of each party's opportunism.

Without law, these parties are not completely helpless. They may anticipate future cooperative projects and will have to take into account the loss of future opportunities if either reneges on their understanding. Moreover, both must worry about their reputations: the firm wants other states to know that it is reliable, and the state wants other firms to know that its representations about the future are trustworthy. But if expectations about future transactions drop, if the gains from opportunism are too great, or if the damage to reputation can be blunted, then both parties can expect the other to exploit its position.

Modern legal theory emphasizes the point that transactors understand this problem up front. In the absence of reliable mechanisms to bolster the credibility of promises, what we should expect to see is not more broken promises, but rather fewer promises made. To stick with our example, if the host state cannot convince the firm that the promise, "You can keep 60 percent if you develop the project" is trustworthy, then the firm will assign a low value to the promise. One of the principal purposes of promise enforcement, in other words, is to benefit promisors by empowering them to make more valuable statements about the contingent future. As a practical matter, this means that fewer projects will go forward absent effective means of reassuring promisees.

Here we see law's function. In some situations, law has the capacity to increase the likelihood that a commitment concerning the contingent future will be honored. How law does this – criminal sanctions, monetary penalties, loss of property or privileges, etc. – is less important than the fact that, when law can impose sanctions that deter promise breaking, it increases the extent and size of potentially valuable investments.

Why do not we use only law to back up our contingent commitments about future states? First, law is costly. Litigation is expensive, time consuming, distracting, and can produce unpleasant surprises. Second, the legal process is fallible. Even the best decision makers make mistakes. Furthermore, facts can be murky, and conditions and factors that matter to one or both parties may be difficult to establish. In legal theory, we distinguish between observable facts – those that the parties to a transaction can detect – and verifiable facts – those that the parties can establish to the satisfaction of a disinterested arbiter at an acceptable cost. Many relevant facts, observable by the parties, may not be verifiable to an arbiter. Finally, the law can be a clumsy instrument for managing the future, applying rules that may be over- or underinclusive for the case at hand.

In any given circumstance, the best means of optimizing the value of commitments involves using some mix of legal rules enforced by formal sanctions and norms enforced informally. The optimal mix will vary with the circumstances of the transactions and the nature of the parties. Legal institutions may be underdeveloped in specific instances, requiring the parties to fall back on norms and informal sanctions. Considerable evidence, however, indicates that giving parties access to a discrete set of rules of sufficient clarity, with credible and independent institutions established to apply these rules and impose appropriate sanctions, enriches most transactional environments.

We have every reason to believe that these generalizations apply to the energy sector. Countries that evolved strong legal institutions in advance of the discovery of energy resources – the United States is one example – have managed to sustain high levels of investment in development and distribution. Some countries with weak institutions – Nigeria is a well-studied example – find it difficult to overcome the temptation to behave opportunistically once parties have committed resources to an energy transaction. Some economists even speak of an "energy curse" to describe the impediments that the discovery of natural riches can throw up for countries not already blessed with strong institutions. As a result, energy assets in institution-poor countries tend to have a lower value compared with those in institution-rich countries.

One readily apparent conclusion to be drawn from this discussion is that resource-rich states have a great incentive to combat the energy curse by pouring resources into law enforcement institutions. To the extent these institutions are seen as possessing strength and integrity, the cost of undertaking energy projects will decline and the extent of such projects will grow. Conversely, failure to address the quality of law enforcement institutions will increase the cost of energy development and distribution.

Note that the quality of law enforcement institutions should not be confused with the quality of legislation. A state may enact model laws dealing with the rights of investors and the state, but such enactments are meaningless without institutions that can see to the implementation of these laws. Courts are critical, but so are the police, the state organs responsible for the enforcement of judgments, and the government agencies that regulate energy projects. I spent the better part of the 1990s working with several governmental and international institutions seeking to improve the legislation affecting economic activity of various transition countries. I came away from this experience deeply frustrated with the apparent willingness of many in the international economic organizations to assume that enacted laws would operate frictionlessly, as if the connection between legal command and social behavior were simple and immediate. Most legal theorists understand that such an assumption does not correspond to observed behavior. The failure to think about law compliance, as opposed to the content of the law, was one of the great failings of the internationally promoted law reform efforts of that period.

II

But which law enforcement institutions affect energy projects? As noted previously, energy development and distribution often involves multiple jurisdictions. To some extent the quality of law enforcement in each affected state will have an impact on the cost of a project. I will focus on the location of the resource to be developed, but obviously the destination of the product, the sources of financial capital, and the jurisdictions through which the product will pass can influence the reliability of the project. If any one of these places has weak institutions, the project can be imperiled.

The quality of the law enforcement institutions at the location of the resource is especially important for four reasons. First, the right to exploit the resource, whether through a license, concession, or other form of permission, turns on the law of the location. The legal stability of this right, including civil law interests in the right, is of obvious significance to everyone involved in the transaction. Second, the local sovereign will have an array of regulatory powers, including environmental protection, labor rules, and, critically, the power to tax. Instability in any regulatory regime drives up the cost of the project. Third, most projects involve the consumption of local inputs, including labor, and these arrangements will be governed by local law. Finally, no matter how international the dispute resolution process used to deal with problems arising from a project, the significance of a judicial or arbitral award rests ultimately on the availability of assets for execution. Even if high-quality institutions undertake dispute resolution processes, the role of law is thwarted unless the outcomes of these processes can produce tangible results. The location of the resource is usually the most likely jurisdiction to have dominion over the assets of critical parties to the transaction, in particular the holder of the right to exploit the resource.

It follows from these observations that the development of logical and fair legal rules governing energy projects is a necessary, but far from sufficient, basis for enabling law to have a positive effect on these undertakings. Nor is it enough to have some effective legal institutions available for the settlement of disputes between investors and host states. The institutions must provide for effective enforcement, which in practice means that credible means for attaching assets must exist in states in which assets belonging to parties to a transaction are located.

This conclusion means in particular that international arbitration institutions, such as the service provided by the International Center for the Settlement of Investment Disputes (ICSID), are an incomplete response to the challenge presented by energy projects. No matter how efficacious the arbitration mechanism, and no matter how respected the arbiters, the prospect of an award will affect the parties' planning behavior, including the decision of whether to go forward with the project, only if they have some confidence that the award will be converted into a financial payment at something like full value. Anticipating that the losing side will hide assets or delay enforcement to frustrate the effectiveness of an award, the parties will undervalue the legal rules that an arbitration process otherwise might bolster.

III

To recap, reasonable legal rules backed up by effective law enforcement institutions can enhance the value of energy development and distribution projects, and weak enforcement institutions subtract from the value of projects. A fundamental aspect of law enforcement institutions' effectiveness is their ability to reassure transactors about the stability of the legal environment in which the project will proceed. For many large-scale energy projects, the institutions of multiple jurisdictions will be involved, and the issue of institutional effectiveness will exist for each jurisdiction. The law enforcement institutions in the jurisdiction where the project is located will be especially important, but other jurisdictions can destabilize the transactional environment.

These broad issues pose several practical legal problems for the transactional lawyer. First, lawyers can attempt to control the number of jurisdictions that might entertain a dispute arising out of the transaction. Second, they can seek to enhance the likelihood that attachable assets will be located in a jurisdiction with effective law enforcement mechanisms. Legal actors in jurisdictions involved in energy projects can support these efforts. None of these efforts can ensure legal stability, but they can produce incremental benefits.

Consider first the question of determining in advance where disputes will be resolved. If the parties prefer effective and disinterested tribunals to those perceived as corrupt or careless, selecting a forum acceptable to the parties involved can increase legal stability. The selection will work, however, only if other jurisdictions respect the choice and refuse to entertain disputes that belong somewhere else.

At one level this is a matter of contract, with choice-of-forum and choice-of-law clauses available for parties who anticipate the problem. The modern trend in the United States has been strongly in the direction of enforcing such clauses, whether the choice involves arbitration or the courts of some other country.[2] But all jurisdictions reserve the right to reject some clauses on public policy grounds, and some regard rights pertaining to energy development as being within this category. For example, Section 50 of the Law on the Subsurface of the Russian Federation, before its amendment in 2000, stipulated that rights created by a license to exploit subsurface resources on Russian territory were subject to the exclusive jurisdiction of Russian courts.[3]

Where parties fail to identify a jurisdiction for considering disputes, they face the risk of exposure to multiple potential resolutions of the same underlying issues. Both the substantive rules and the nature of the law enforcement institutions can vary significantly. Under these circumstances, the question of stability can present itself in unpredictable ways.

One issue that has emerged in recent U.S. litigation involves the application of the *forum non conveniens* doctrine to disputes over rights to Russian natural resources, including energy development projects. The disputes are interesting because many lawyers, especially those trained in the United States, regard the U.S. litigation system as the epitome of developed, predictable, stable law. This view, however, ignores the unsettling effect of the civil jury system, the unique pretrial discovery practices of U.S. courts, the class action device, contingent fee arrangements, and the availability of punitive damages in private litigation. These facets of civil litigation, which U.S. lawyers take for granted, can convert otherwise outlandish claims into lawsuits that pose a significant financial risk to the defendant. Moreover, the nature of U.S. civil litigation makes precise *ex ante* valuation of such suits extremely difficult.[4] A legal doctrine, such as *forum non conveniens*, that can pretermit a U.S. lawsuit responds to this particular form of instability.

As I have observed elsewhere, the United States uses the *forum non conveniens* doctrine as much as it does largely because U.S. rules of court jurisdiction are so liberal.[5] Because jurisdictional rules do not screen out lawsuits that should not be brought in U.S. courts, another tool is needed, one that gives courts greater discretion to pick and choose. The doctrine turns on two categories of inquiries: whether, balancing all relevant factors, another forum would provide a clearly superior venue for the suit; and whether the alternative forum is "adequate."

I have worked as a Russian law expert in several lawsuits involving disputes over the ownership of interests in Russian natural resource projects, one of which involved oil. In each case, Russian courts and, in some instances, international arbitration bodies previously had heard versions of the claims presented by the U.S. plaintiffs. In each case, the plaintiffs alleged that Russian courts have participated in a concerted plan to deprive them of their interests, and that the purported corruption of the Russian judicial system made that jurisdiction an inadequate forum to hear the plaintiffs' claims. Finally, in each case the U.S. court rejected the plaintiffs' assertions and applied the *forum non conveniens* doctrine in favor of Russian courts.[6]

For present purposes, the holding of these cases is less important than the underlying logic of the decisions. Judge Koetl, writing in the first of the cases, observed:

> The contracts that are at issue in this case demonstrate that the plaintiffs should not have expected that any of their disputes would be litigated in the United States. This makes sense in view of the fact that the contracts were for services to be performed in Russia.[7]

He tied the application of the doctrine to the reasonable expectations of parties to a transaction, and inferred those expectations from the nature of the transaction. Lawsuits arising out of projects undertaken in a particular country, he suggests, ought to be heard by courts in that country, unless the parties expressly agree otherwise. This reasoning, if more widely embraced by other courts, will reduce the legal instability associated with energy development and distribution undertakings.

One also should note, however, that a "reasonable expectations" approach applies principally to persons who choose to enter into a transaction, in particular investors. Its limits can be seen by considering another class of disputes, where plaintiffs assert that energy projects have caused injuries to third parties, such as local indigenous groups. Allegations typically depict a major company as aware of or otherwise involved in repressive actions employed by the local authorities to remove obstacles posed by the local population to extraction and distribution facilities. U.S. courts have upheld the validity of these legal theories, although no court has rendered a final judgment in any of these disputes.[8]

These claims typically involve alleged violations of human rights law. The United States is unique in allowing private suits for damages based on violations of customary human rights law, and some of the legal theories used to sustain the practice are under fire.[9] But even if the legal basis for claiming compensation changes, and plaintiffs find themselves relegated to U.S. state courts rather than the federal courts, it seems reasonable to expect an ongoing legal risk from the United States to the extent that local authorities engage in objectionable practices and any party to a transaction is subject to the power of the U.S. judiciary.

Up to now I have explored how transactors can try to stabilize the legal consequences of their actions by selecting one and only one jurisdiction for the resolution of their disputes, and how states can approximate that ideal by surrendering jurisdiction to those places that parties reasonably would have anticipated to be the appropriate forum for their lawsuits. When it comes to the enforcement of judgments and arbitral awards, however, a different dynamic applies. For dispute resolution to be effective, implementation of a judgment – especially a monetary payment – must be as frictionless as possible. To that end, the parties must have assets available in jurisdictions that have effective collection systems. Multiplying the number of jurisdictions and the amount of assets available increases the likelihood that awards will be converted into tangible sanctions without any offsetting costs.

With these considerations in mind, parties to energy projects often require in advance that revenues from the sale of the product pass through a designated bank account in a jurisdiction known for effective collection institutions, such as New York. Payments to the various stakeholders in the project are disbursed out of this account, but any money passing through can be used to satisfy judgments. In theory, the account functions as something like a guaranty against the satisfaction of proper judicial judgments or arbitral awards.

Implementing this concept, however, can be complicated. The leading case arose out of a project undertaken by a power company formed by U.S. investors and the Indonesian state-owned entity Pertamina. After the Indonesian financial crisis intervened, the government canceled the power company's energy development contracts, and in response the company obtained a substantial arbitral award. To enforce the award, the company searched for Pertamina assets in the United States. In accordance with the terms of numerous production sharing contracts for the development of liquified natural gas, Pertamina channels its sales revenues through a New York bank account. The power company sought to collect out of that account. But the U.S. court determined that Pertamina did not hold the lion's share of these funds on its own behalf. Rather it acted as the agent of the Indonesian state, which was entitled to royalties. Neither the production sharing contracts nor the terms of the bank account contained language that would override Indonesian law, which treated the state as the owner of the royalties from the moment that Pertamina collected the money.[10]

The power company in the Pertamina dispute was not a party to any of the production sharing contracts in question, and its contracts with Pertamina made no mention of the New York bank account. In retrospect, it had missed a critical opportunity to enhance the quality of its legal rights. Moreover, one wonders why the parties to the production sharing agreements had gone to the trouble of routing money through a U.S. bank but not of limiting the Indonesian state's claim on the revenue to one of equality with all other creditors. One would think that, if the parties had anticipated this issue, the Indonesian government might have agreed to accept the role of creditor, rather than as owner of the funds in the bank accounts, as a way of reassuring the other parties and thereby lowering the costs of the projects.

IV

Each of the discrete legal issues discussed in this paper – the enforceability of choice-of-forum and choice-of-law clauses, the applicability of *forum non conveniens* doctrine, and the specification of ownership rights of attachable assets – illustrates ways in which transactional lawyers can try to increase law's contribution to the value and feasibility of energy projects. I focus on these examples because they involve specific features of the legal system – the availability of dispute resolution and the enforcement of the decisions that dispute resolution bodies reach – rather than the specification of the primary rights and responsibilities of the parties. They also are incremental. Their limits reflect what law can and cannot do.

Law cannot protect against unanticipated changes in economic fundamentals, such as sharp swings in energy prices or consumer demand. Law cannot guard against political reversals or social turmoil. When the Asian financial crisis broke over Indonesia in the late 1990s, energy projects were unwound or disrupted, whatever the strength of the legal guarantees that the participants enjoyed.

Energy projects in Russia also illuminate the limits of legal change. Ongoing uncertainty in Russia over the validity and durability of the legal structures created during the 1990s for holding and developing energy rights has limited that industry's access to needed capital. Because this uncertainty is caused by the status of existing law, rather than its content, new laws cannot help. Measures to bolster legal institutions might alleviate these concerns, but outsiders may have difficulty distinguishing reform efforts from governmental interference in the operation of the legal system. Ironically, countries that have acquired a reputation for legal instability have less freedom to undertake innovation and change in their legal systems than do states with well-established institutions.

The challenge for lawyers working on energy projects thus is clear: They should do what they can to strengthen the legal protection of their clients' rights, and advise their clients of the limits of that protection. As citizens, they can advocate the strengthening of law enforcement institutions and publicize the costs that weak institutions produce. They cannot solve all problems, but they can help some.

Notes

1. For further work by theorists, see R.E. Scott, "A Theory of Self-Enforcing Indefinite Agreements," 103 *Columbia Law Review*, No. (2003); A. Schwartz & R.E. Scott, "Contract Theory and the Limits of Contract Law," 113 *Yale Law Journal*, No. (2003).

2. *Vimar Seguros y Reaseguros, S.A.* v. *M/V Sky Reefer*, 515 U.S. 528 (1995); *Mitsubishi Motors Corp.* v. *Soler Chrysler-Plymouth, Inc.*, 473 U.S. 614 (1985); *The Bremen* v. *Zapata Off-Shore Co.*, 407 U.S. 1 (1972).

3. Federal Law No. 20-FZ of January 2, 2000 on Introduction of Amendments and Addenda into the Law of the Russian Federation on Subsurface, Art. 1, par. 17.

4. P.B. Stephan, "Redistributive Litigation – Judicial Innovation, Private Expectations, and the Shadow of International Law, 88 *Virginia Law Review*, No. 4 (2002), 862-65.

5. P.B. Stephan, "A Becoming Modesty – U.S. Litigation in the Mirror of International Law, 52 *DePaul Law Review*, No. 2 (2002), 634-35.

6. *Norex Petroleum Ltd.* v. *Access Industries Inc.*, 304 F. Supp. 2nd 570 (S.D.N.Y. 2004); *Base Metal Trading, S.A.* v. *Russian Aluminum*, 253 F. Supp. 2nd 681 (S.D.N.Y. 2003), affirmed, ___ F.3rd __ (2nd Cir. 2004) (unpublished). For related authority, see *Monegasque de Reassurances S.A.M.* v. *NAK Naftogaz of Ukraine*, 311 F.3d 488 (2nd Cir. 2002).

7. *Base Metal Trading, S.A.* v. *Russian Aluminum*, at 696-97.

8. *John Doe I* v. *Unocal Corp.*, ___ F.3rd ___ (9th Cir. 2002) (unpublished pending rehearing en banc); *Wiwa* v. *Royal Dutch Petroleum Co.*, 226 F. 3rd 88 (2nd Cir. 2000).

9. *Sosa* v. *Alvarez-Machain*, 542 U.S. ___ (2004).

10. *Karaha Bodas Co., L.L.C.* v. *Perusahaan Pertambangan Minyak Dan Gas Bumi Negara*, 313 F.3rd 70 (2nd Cir. 2002).

A Conflict of Interest for Russia: Offshore Oil vs. the Problems of Environmental Regulation

Richard N. Dean[*] *and Michael P. Barry*[†]

Abstract: *This survey paper will examine the problems Russia faces in regulating energy exploration. On the one hand, Russia possesses massive oil and gas reserves, the sale of which constitutes a major portion of the developing Russian economy. Europe depends on Russia's gas supplies, Asian demand is growing, and Russians themselves will consume more and more energy as their economy grows. At the same time, however, energy exploration poses great risks to the environment, including tanker spills, pipeline breaks, discharges, seismic testing, and other risks. This paper will examine these risks, and discuss the current regulatory regime Russia uses to manage them.*

I. Introduction

Russia accounts for approximately 5-6 percent of the world's proven oil reserves and about one-third of the world's natural gas reserves. With a similarly large endowment of coal, Russia is a country quite dependent on energy resources. Energy accounts for about 30 percent of Russia's GDP, and in 2003, the export of crude oil, oil products, and natural gas accounted for 55 percent of Russia's goods exports and a full 17 percent

[*] Richard Dean is a partner in the international law firm of Coudert Brothers LLP. He heads the firm's Russia and Central Asia Practice and focuses on corporate transactions, including mergers and acquisitions, foreign direct investment, and joint ventures, and compliance issues, including the Foreign Corrupt Practices Act and anti-money laundering legislation. Mr. Dean opened Coudert Brothers' Moscow office in February 1988 and was a resident there until relocating to Washington, D.C., in 1991.

[†] Michael Barry is a PhD economist at Coudert Brothers LLP. Before joining Coudert Brothers, he worked as an international economist at the U.S. International Trade Commission. Dr. Barry also served as a consultant to the U.S. Department of Treasury and the Russian Ministry of Finance in Moscow. He wrote his doctoral thesis on the Russian economy, and is now studying law at the American University in Washington, D.C.

of GDP.[1] Russia's economy has benefited significantly from the high international oil prices since 1999, and stands only to gain more from large-scale investment projects in offshore oil and gas as well as other energy sectors.

On the other hand, years of neglect under the Soviet Union has left Russia with significant environmental problems. Though by international standards the former Soviet Union had rather advanced legislation on environmental protection, a lack of enforcement undermined its effectiveness. The collapse of the Soviet Union and the significant contraction of the economy actually reduced pollution in Russia by reducing energy consumption and carbon emissions. But as Russia emerges from the collapse of the USSR and the 1998 economic crisis, it is enjoying rapid GDP and increased investment in oil and gas.

The conflict Russia faces is not particularly unique. Many countries struggle with the need for foreign investment and development of energy infrastructure while at the same time hoping to manage and protect their resources and environment. This paper will examine this dilemma facing Russia and other countries, and review some of the methods used to balance the conflicting incentives faced by energy-producing countries. It might be said that *the greatest problem in regulating the development and production of offshore hydrocarbon deposits in Russia is simply the strong market forces pushing for the development of this very sector.*

II. PUSH FACTORS: WHY OIL AND GAS?

It is not difficult to see why Russia sees incentives to develop its energy sector. These incentives can be split into two groups: (1) supply advantages; and (2) demand opportunities. A third, less pleasant incentive might be called momentum or inertia: Because Russia is already so dependent on energy, the Russian economy faces significant challenges in trying to diversify into anything else.

a. Supply Side Advantages

First are the easy supply advantages. Russia holds the world's largest natural gas reserves, with 1,680 trillion cubic feet – more than twice the reserves of the next largest country, Iran.[2] In oil, Russia has proven oil reserves of 60 billion barrels, most of which are located in Western Siberia between the Ural Mountains and the Central Siberian Plateau.[3]

Sometimes called the "Russian Core," in 1988 the region's production peaked at 12.5 million bbl/d. Soviet oil production surpassed even that of Saudi Arabia, but production plummeted after depletion of the largest fields and then the collapse of the Soviet Union. Production fell 44 percent between 1988 and 1994. The government's continued subsidization of oil and unrealistically low domestic oil prices meant little interest in investment and the collapse of Russian oil.[4]

But several factors have contributed to a significant rebound, including high energy prices, devaluation of the rouble following the 1998 financial crisis, and the inflow of investment and new technology from abroad. According to forecasts of the Russian Energy Ministry, oil production will average 8.8 million bbl/d in 2004.[5]

Enormous offshore reserves are located in the Russian Far East. This region shares a 1,300 mile border with China, and also borders Mongolia, North Korea, the Sea of Japan, the Sea of Okhotsk, the Pacific Ocean, and the Arctic Sea. Though the region is one of the least developed areas of the Russian Federation, it produces almost all of the country's diamonds and tin, and more than half of Russia's gold and fish. The energy sector in the region includes oil, natural gas, coal, and nuclear power.[6] Approximately 60 percent of the region's oil reserves are on land, and the rest is located on the adjacent continental shelf of Sakhalin Island in the Sea of Okhotsk.[7]

The Sakhalin oil and gas projects, which have grown into one of the largest direct investments in the Russian economy, involve the participation of a large number of Russian and foreign energy companies, suppliers, manufacturers, construction and design companies and accompanying service providers. The projects were the first in Russia to employ production sharing agreements (PSAs) (to be discussed in detail in this paper). PSAs attracted a large amount of foreign investment by providing favorable tax treatment (substitution of the majority of federal and local taxes with product sharing) and guaranties of stability for the whole period of the project implementation activities (30-40 years). This allowed for the successful development of these large-scale projects in harsh environmental and climate conditions.[8]

Sakhalin Island, a former penal colony located off Russia's Pacific coast, is home to five oil and gas projects, each operated by an international consortium. The five projects are currently in different stages of development.[9] The main attractions of the projects include: (1) the enormous energy resources of the region; and (2) the proximity to Asian markets, including Japan, China, and Korea.[10]

The Sakhalin I project is led by Exxon Neftegaz, in conjunction with SODECO, ONGC Videsh, Sakhalinmorneftegaz, and RN Astra. The consortium members started drilling in May 2003 and expect preliminary oil production of 250,000 bbl/d in 2005. The project's oil output will be piped westward to the Russian port of DeKastri and pumped into the existing pipeline system. Sakhalin I's natural gas is expected to be piped southward to Japan via a proposed pipeline. The partners are planning for natural gas exports to Japan to begin in 2008.[11]

Sakhalin II is being developed by Shell, Mitsubishi, and Mitsui. The Sakhalin II project involves the development of Russia's first liquefied natural gas (LNG) facility, to be built on the southern tip of Sakhalin Island, near the town of Prigorodnoye. The operation will have a capacity of 9.6 million tons per year, with exports to begin in 2007. According to press reports, the Sakhalin II partners have already secured sales contracts with three Japanese utilities worth 2.8 million tons per year, or approximately 30 percent of the plant's output, for roughly 20 years.[12]

Sakhalin III involved the development of two independent projects: the Kirinskiy Perspective Block (roughly 687 million tons of oil and condensate, 873 billion cub. m. of natural gas) and the Vostochno-Odoptinskiy & Ayashskiy Perspective Block (roughly 160 million tons of oil and condensate, 67 billion cub. m. of natural gas). Both projects require geological and exploration research to determine the exact volume of recoverable oil reserves and exact location of oil and gas deposits. The operator and investor of the first project is PegaStar (USA); its founders are Mobil Russia Ventures Inc, Texaco Exploration Sakhalin Inc, NK Rosneft and Rosneft-Sakhalinmorneftegas JSC. The investors in the second project are Exxon Neftegas Ltd, NK Rosneft and Rosneft-Sakhalinmorneftegas JSC.[13]

In early 2004, however, the Russian Cabinet annulled the tender for the Far East Sakhalin III oil fields won by a consortium led by Exxon Mobil Corp. and ChevronTexaco Corp. and stated that no license will be issued for its development. The consortium, which also included Russia's state-owned Rosneft oil company, won the tender in 1993, but a mineral license was never issued in the absence of a legal framework for production-sharing agreements.[14]

While the legal environment for energy investment appears to be changing in Russia, the vast reserves in the Sakhalin region will continue to attract investors. Sakhalin, roughly equal in area to the state of Maine, but with fewer than half as many residents as Maine's 1.2 million people, is on track to surpass Moscow as the country's leading destination for foreign investment.[15]

b. Demand Side Opportunities

Abundant supply is not the only incentive for Russia to develop its oil and gas fields. There are at least three major demand factors, including: (1) Russia's own economic development and rapid growth; (2) Europe's dependency on Russian energy supplies; and (3) East Asia's growing energy demands and links to Russian energy supplies.

First, after a decade of stagnation, Russia now has a growing economy. Five years after defaulting on its debts and devaluing the rouble, Russia has made great strides to protect itself from another financial crisis.[16] Russia's real GDP growth over the past five years has averaged an impressive 6.6 percent per year – a rate few other countries have matched. Stimulated by the 1998-99 devaluation and high energy prices, idle capacity in many industrial sectors was brought to life.[17] Privately owned factories across Russia are working at full capacity to satisfy domestic demand and driving Russian economic growth – the strongest in nearly a century. In 2003 the Russian economy grew by 7.3 percent, bringing cumulative growth since the 1998 financial crisis to 38 percent. Inflation has been gradually declining and in 2003 reached 12 percent. The capitalization of the Russian market has gone up fivefold since 1999 to $250 billion.[18]

And as Russia's economy grows, so will its own energy needs. A good example is China, where GDP growth consistently exceeds 8 percent annually. The economic boom in China is creating shortages in goods and commodities around the world, raising prices and sparking investment in many metals and commodities. While Russia has a smaller population and has not been the recipient of the record foreign investment that China has, the economic development of Russia will require more and more energy.

The second demand factor is The European Union. The EU is increasingly dependent on energy from Russia, particularly natural gas. The European Commission predicts that given present trends, the EU's current 50 percent dependence on imported energy will rise to 70 percent by 2020.[19] Natural gas is currently the major topic of Russia's WTO accession negotiations. Russian officials believe the EU may be willing to soften its critical stance on one of the key issues of WTO talks – Russia's system of energy pricing – in exchange for a pledge of open access to gas pipelines for competitors to the state-backed Gazprom.[20]

A 1997 EU directive opened the EU natural gas market to competition. The legislation first took effect in 2000, when most member States were to open an initial 20 percent of their markets to competition, starting with the largest consumers, rising to 28 percent in 2003. Most member States have exceeded the minimum requirement, with several having fully liberalized their markets. On average 80 percent of EU gas demand is now open to competition.[21] Russia hopes to gain from this liberalization.

The third major demand side incentive is the Asian market. After the oil crises in the 1970s, Japan invested heavily in energy conservation measures, and as a result, industrial demand for energy remained almost flat in the 1980s, despite high rates of economic growth. Energy consumption by industry fell in the early 1990s, but consumption by households and the transport sector has continued to rise. The growing use of cars and of air conditioners in the home in the hottest months of the year, between June and September, are important factors contributing to this development.[22] Though Japan has accumulated a significant strategic stockpile of oil, it is the world's second leading importer of oil. Almost 100 percent of both oil and natural gas is imported into Japan.

Japanese officials have taken note of the country's growing reliance on oil from the Persian Gulf – up to 86 percent of imports in 2003 from about 70 percent at the time of the 1991 Persian Gulf War. In April 2003, Japan mounted a lobbying campaign to persuade Russia to build a 2,500-mile pipeline to carry a million barrels of oil a day from eastern Siberia to Nakhodka, a Russian port facing Japan. Tokyo even offered to finance the entire $5 billion pipeline, which could meet nearly one-quarter of Japan's oil needs.[23] The Russian government instead decided to back a shorter and cheaper rival pipeline proposal to supply northeastern China. Though this oil would go to China, the goal of reaching Japan remains.

As for China, with oil consumption expected to double this decade, the country will soon overtake Japan as the world's No. 2 consumer after the United States, and could easily swallow all of Russia's East Siberian oil exports, leaving little need to extend the pipeline.[24]

According to the China National Offshore Oil Corporation (CNOOC), China will provide a $60 billion market share of LNG to foreign enterprises by 2020. The CNOOC reports that China's output of natural gas cannot meet the rising demand for clean energy in the country in the next 15 to 20 years. Demand for LNG in the next 15 years will rise by 12 percent annually, and it is predicted that the annual consumption of natural gas in China will reach 160 to 210 billion cubic meters by 2020. CNOOC says that 49 percent of China's natural gas needs will be met by imports, with 39 percent from LNG imports and possibly 10 percent provided by the pipeline project from Russia and Central Asian nations.[25]

Whether China, Japan, or other countries like Korea become the next destination for Russian oil and gas, Russia stands to gain. Russia has enormous supplies, and much of these supplies are located in close proximity to China, Japan, and the other potential importing countries. At one point, Sakhalin Island is as close as 50 miles from the northern coast of Japan's Hokkaido. The commercial director of Sakhalin Energy Investment Company Ltd. claims that Sakhalin will soon supply 15 percent of all gas going into Tokyo, and that half of the consortium's gas output will go to Japan.[26]

c. Russia's "Inertia" in Energy

A final incentive for Russia to further develop its energy sector is simply self-preservation: The Russian economy depends on energy sales. According to a recent World Bank study, the Russian economy is nearly three times more dependent on oil and gas than official statistics indicate, making the country much more vulnerable to oil price swings than previously thought.[27] The State Statistics Committee does not properly account for the tax avoidance schemes used by oil and gas companies, the bank concluded in its annual report on the Russian economy. But if it did, the numbers would show that oil and gas accounts not for 9 percent of gross domestic product, but 25 percent, and that services account for just 35 percent of GDP, not 55 percent.[28] The new figures show that the oil and gas sector, despite employing less than 1 percent of the workforce, makes up half of all industrial production.[29] Oil and other natural resources account for 80 percent of all exports, and taxes on the oil and gas sector generate two-fifths of all government revenues.[30]

According to research by Economist Intelligence Unit, the elasticity of real GDP to the oil price in Russia is estimated at 0.08. This means that a 10 percent change in the oil price leads to a change in real GDP growth by 0.8 percentage points. In other words, at present levels, a $1 drop (rise) in the oil price leads to an increase (decline) in real GDP growth of about 0.4 percentage points.[31] In simpler terms, the Russian economy is very dependent on its oil sales, and growth is very sensitive to the oil price. While policymakers in the Kremlin talk much of measures to diversify the economy from energy, the present dominance of oil and gas is bound to slow any changes.

An additional factor is what some economists call the "Dutch Disease." Named after the exchange rate phenomenon encountered when oil deposits were discovered near the Netherlands, the Dutch Disease describes a real appreciation of domestic currency caused by natural resource extraction. In a seminal 1995 paper examining the performance of a large number of countries between 1970 and 1990, Jeffrey Sachs and Andrew Warner demonstrated that the more a country is dependent on the export of natural resources, the lower its economic growth. The conclusion was that a large resource endowment leads to poor economic policies and the so-called Dutch Disease, when abundant resource exports boost the exchange rate excessively, rendering other industries uncompetitive.

The Dutch Disease is arguably a true phenomenon in Russia, where exports of oil and gas have been accompanied by a real appreciation of the Russian currency and stagnant industrial growth. Such a phenomenon is pernicious to the Kremlin's attempts to diversify the economy because, essentially, strong energy exports make Russian manufacturing uncompetitive.

III. PULL FACTORS: ENVIRONMENTAL COSTS OF OIL AND GAS EXPLORATION

The other side of the equation is potential harm to the environment. The balance between attracting investors to develop oil and gas fields and the growing concern over protecting the environment is a question that currently confronts many parts of the globe. The governments of Brazil and Venezuela are rapidly developing both their onshore and offshore oil and gas reserves, as are other countries in Latin America, Africa, and the former Soviet Union.

There is vast literature that examines the economic impact of offshore oil and other energy exploration. The impacts include both ecological and commercial effects. Ecological effects include changes to the topography of the sea floor, destruction of marine life and its habitat, and damage to the equilibrium of the ecosystem. While oil tanker accidents and large oil spills possibly attract greater attention, far more common are the small spills and intentional discharges that occur daily.

Economic losses include the loss of fish stock for commercial fishermen and physical damage to fishing equipment. Spilled oil disrupts gestation and development of new fish, kills and harms marine life, and injures the habitat of the fish stock. The resulting economic losses are often highly concentrated in countries and cities in which fishing is a dominant sector. This section will briefly outline environmental issues in offshore oil and issues of particular concern to Russia.

a. Tanker Spills

Environmentalists have two major concerns regarding tanker transport: (1) the risk of tanker spills; and (2) the inadequacy of response capabilities in the event of a catastrophe. According to one case study of Sakhalin Island oil, tanker transportation accounts for over 50 percent of the world's oil spills.[32] Weather and traffic in the Sea of Okhotsk and other sites increase the likelihood of an accident.[33] The Sea of Okhotsk is subject to violent storms, tsunami, sea currents, ice sheers, and fog, which decrease visibility. Storage facilities with up to 140,000 tons of oil and typical tankers with a capacity of up to 90,000 tons are thus at greater risk of collisions and spills. In addition, many of the tankers in use do not use "double-hull" technology, meaning even less protection from catastrophe.[34] Finally, the success of development has produced greater traffic of tankers and other ships in the region between Sakhalin, the Russian mainland, and Japan.[35]

A second concern of environmentalists is that Sakhalin Island and Russia do not have adequate spill response mechanisms and plans for mitigating damage when an accident occurs. Two issues emerge: (a) Russia's lack of an oil response industry; and (b) inadequate financial commitments to pay for accident response. The Russian government has little equipment designed for oil spill clean-up, instead relying heavily on contingency plans that would require up to 10 days delay before actual clean-up could begin.[36]

The second issue is the financing of any clean-up. According to one study, the financial liability of the tankers used by an oil company operating offshore of Sakhalin Island measures $81 million, which is much less than, for example, the $2 billion cost of cleaning up the 40,000 tons of oil spilled by the Exxon Valdez in Alaska.[37] The company reports that the cost of clean-up would be paid for by insurance policies or by international treaties. But environmentalists say that insurance is not available for certain accidents such as acts of God or errors by personnel. Further, environmentalists believe that such oil companies could simply cite limited legal obligations and refuse to pay any sum greater than their fixed investment cost in the project.[38]

Some environmental groups suggest that energy companies should be required to assume full liability for any spills, and that multilateral investment institutions such as the World Bank or EBRD should guarantee the liability.[39]

b. Discharge of Drilling Muds and Cuttings

A second source of pollution involves waste discharge. During offshore drilling, water-based drilling muds and cuttings are discharged overboard, whereas during production, the major discharge is "produced" water. In addition, treated sanitary and domestic wastewaters, deck drainage, and miscellaneous wastes, such as ballast waters, may be discharged at some point in the operation.[40]

Drilling fluids or "muds" are used during drilling operations to remove rock cuttings, lubricate and cool the drill bit and string, control the formation pressure, and seal the well. About 95 percent of all drilling mud comprises water, clay, and barite. The rest are additives, such as starch, lime, soda ash, or sodium bicarbonate. In the United States, oil-based muds require special permits and are used for high-temperature wells, deep holes, or when hole stabilization poses a problem.[41]

Some environmentalists criticize companies that produce drilling fluids. According to one group, individual drilling companies have devised proprietary and secret formulations ("mud recipes") to deal with specific types of drilling jobs. One of the problems in studying the effects of drilling waste discharges is that the drilling fluids are made from a range of over 1,000 ingredients – many of them known, confusingly, by different trade-names, generic descriptions, chemical formulae and regional or industry slang words.[42] Environmentalists claim that the exact composition of drilling fluids, and thus their harmful effects on the environments, is not fully understood.

Produced water is mainly salty water trapped in the reservoir rock and brought up along with oil or gas during production. It can contain chemicals added downhole during production. These waters exist under high pressures and temperatures, and usually contain oil and metals. Because of this, they must be treated prior to being discharged overboard. As with drilling muds, following treatment, they must be tested for toxicity and cannot exceed set discharge rates. In some areas, produced water is likely to contain naturally occurring radioactive material.[43]

There are several technologies in use today that can lessen the effects of waste discharge. First, energy companies are urged to use less toxic alternatives. A second technology is cleaning onboard prior to discharge. More efficient equipment has been developed to separate solids from drilling fluids after use. This equipment can improve the separation of mud from cuttings and oil from produced water. Further, technology exists that can largely remove the residual pollutants in cuttings and, particularly, in produced water. Though technically feasible, these techniques are still prohibitively expensive to many projects. A third technology is to simply ship wastes ashore for recycling, landfill and/or incineration.[44] While this technique is used in many projects, many environmentalists claim this merely moves the pollution from the sea to the land and air.

A final technique for minimizing waste discharge is "re-injection."[45] Cuttings re-injection (CRI) is a waste disposal technique where drill cuttings and other oilfield wastes are mixed into a slurry with water and pumped at high pressure down an injection well. Sometimes it is necessary to grind up the particles in the slurry to make them finer. The hydraulic pressure can also be used to break open layers in the rock to make subsequent injection easier and to contain the wastes in a defined area. A name commonly used in the United States for the technique is "slurry fracture injection." In the United States, oil companies are moving toward "zero discharge" standards in which nearly all drilling muds and cuttings are re-injected into the oil field. Most oil industry observers agree that re-injection is now the "best available technology."[46] Though companies in the Sakhalin Island region have also employed the technique, environmentalists are urging more universal acceptance.[47]

c. Pipelines

A third source of risk is transport of oil by pipelines. The third largest oil spill in history occurred after the failure of a Russian pipeline near the town of Usinsk in the Komi Republic of northern Russia.[48] It is estimated that between 14,000 tons and 100,000 tons of oil or more was released into the northern Russian environment range. The official figure from the Russian Ministry of Environment is 60,000 tons, which exceeds the Exxon Valdez oil spill, though the Komi spill did not receive the same level of worldwide attention.[49] The spill was caused by old pipelines that began leaking long before the eventual spill. The ruptured pipeline was 150 km long and was constructed in 1975. It was estimated to be beyond operational guarantees starting in 1990.[50] The rupture resulted in severe oil contamination of the Komi region. The spill reached the Kolva River, a tributary of the Pechora River, which falls into the Barents Sea.[51] The World Bank assisted cleanup efforts with a $100 million loan, but environmentalists cite the case as proof of a decaying Russian oil infrastructure.[52]

The same pipeline risks affect offshore oil projects. The European Oil Company Organisation for Environment, Health and Safety (CONCAWE) considers the risk of pipeline ruptures as a function of several factors, including mechanical problems, corrosion, third-party activities, operational variables, and climate or environmental changes.[53] Mechanical failure can occur either during construction as a result of negligence or the utilization of poor quality pipeline materials, or as a result of structural deterioration of the pipeline material (wear and tear) due to age. Generally, the older the pipeline, the greater the probability of mechanical failure. Corrosion of pipelines occurs both internally and externally, and often results from the lack of anticorrosion coatings. Operational errors include both system failure and human errors, including lack of adequate maintenance. Third-party activities can damage pipelines, whether accidental, malicious (such as sabotage), incidental, or by acts of war. Natural hazards to pipelines include ground surface subsidence, river flooding, wind erosion, and rapid changes in temperatures.[54]

There were approximately 84,000 kilometers of pipeline in the former Soviet Union as of 1998. About 90 percent of this pipeline has a diameter of greater than 20 inches. About 64,000 pipeline kilometers, or 76 percent of the total, are located in Russia. A large number of the oil pipelines in Russia began operation in the 1960s and 1970s, and as of the year 2000, pipelines older than 20 years accounted for 41 percent of the total Russian pipeline network length.[55] There is an increase in operational hazards with older pipelines, which could be due to the aging of the steel used in pipeline manufacturing. Aging of steel leads to undesirable changes in its properties, including a decrease in plasticity.[56]

d. Seismic Surveys

Another environmental risk is the effect of high-energy seismic surveys on marine mammals. Seismic surveys are a component of many information gathering efforts conducted in many parts of the offshore and coastal waters. Seismic surveys use very short, high-energy impulses of sound directed downwards into the seafloor, with some energy being reflected back. There is concern that these pulses, which lie toward the lower end of the range of human hearing, may adversely affect nearby marine mammals, particularly endangered species.[57]

Modern seismic surveys commonly consist of boats towing either single airguns or an array of airguns that produce loud, low-frequency impulsive sounds at frequent and regular intervals, in zig-zag formations across the ocean. The intense sounds produced by the airguns then reflect back to the vessel to provide information about oil and gas deposits that may lie beneath the seabed.[58] Although the sound pulses are aimed down to the ocean floor, there is often significant horizontal propagation, which can be detected more than 100 km away.[59] Environmentalists believe that the low frequencies negatively impact mammals that communicate using infrasonics, such as whales. However, certain higher frequencies are also feared to affect whales, dolphins, and porpoises.

e. Grey Whales and Other Wildlife in Russia

In Russia, much of the vast oil and gas reserves of the Far East happen to lie in an especially sensitive environment. The Sea of Okhotsk is one of the world's most biologically rich seas. The people of the Russian Far East depend on the large stocks of crab, shrimp, pollock and other seafood products in the Sea of Okhotsk. The coastlines of the Sea of Okhotsk still provide spawning grounds for healthy, wild Pacific salmon runs that are in decline in other parts of the North Pacific. The waters near northeastern Sakhalin provide a habitat for the endangered Western Grey Whale. Sakhalin's economy has long been based on its fisheries, and anything that negatively impacts Sakhalin's fisheries will negatively impact the region's economy and local employment. Indeed, the Sea of Okhotsk provides more than half of Russia's yearly fish catch.[60]

Julia LeMense Huff provides a detailed discussion of the special sensitivity of the Western North Pacific grey whale, which resides in the Sea of Okhotsk.[61] The grey whale is listed as an endangered species by both the Union for the Conservation of Nature (IUCN) and the Russian Federation Red Book. Though the exact whale population is not known, estimates put the number at approximately 100. The grey whale was hunted to extinction in the North Atlantic Ocean over 150 years ago, so the focus on their survival has shifted to the North Pacific Ocean. The grey whales migrate between an unknown mating and calving ground and their primary feeding ground in the shallow shelf waters and offshore banks on the coast of Sakhalin Island in the west central Sea of Okhotsk, in an area that overlaps with the Odoptu oil field that is part of the Sakhalin-1 Project.[62]

In response to concerns over the grey whale, Sakhalin Energy Investment Company, the operator of the Sakhalin-2 Project, recently redrew its timetable for laying a pipeline across the sea floor to mitigate the negative effects of construction work on the Korean-Okhotsk population of the whale. The whales live off the Korean Peninsula in the winter, but migrate to Sakhalin in the summer. One of the main feeding areas is the so-called Piltun-Astokhskoye oil and gas deposit, where the Molikpaq sea platform has been established. Construction on an underwater pipeline from the platform was due to start in 2004, and the pipe was to be laid at the bottom of a ditch dug across the sea floor. Such construction would require use of more powerful suction dredges, whose noise would frighten off the grey whales. To avoid this, Sakhalin Energy decided to carry out additional acoustic research to find out how to minimize the damage that dredging vessels can do to whales. Until then, the underwater pipeline will be laid where no whales have been spotted.[63]

f. Foreign Energy Companies Making an Effort

Foreign energy companies are well aware of the environmental risks and the allegations of environmentalists. Foreign energy companies publicly announce their environmental polices and take their commitments seriously. For example, ExxonMobil publishes the following as its environmental policy:[64]

- comply with all applicable environmental laws and regulations and apply responsible standards where laws and regulations do not exist;

- encourage concern and respect for the environment, emphasize every employee's responsibility in environmental performance, and foster appropriate operating practices and training;
- work with government and industry groups to foster timely development of effective environmental laws and regulations based on sound science and considering risks, costs, and benefits, including effects on energy and product supply;
- manage its business with the goal of preventing incidents and of controlling emissions and wastes to below harmful levels; design, operate, and maintain facilities to this end;
- respond quickly and effectively to incidents resulting from its operations, in co-operation with industry organizations and authorized government agencies; conduct and support research to improve understanding of the impact of its business on the environment, to improve methods of environmental protection, and to enhance its capability to make operations and products compatible with the environment;
- communicate with the public on environmental matters and share its experience with others to facilitate improvements in industry performance; and
- undertake appropriate reviews and evaluations of its operations to measure progress and to foster compliance with this policy.

In the United States, energy companies respond to the concerns of environmentalists by showing that offshore oil and gas production is both safe and important to the economy. One analyst points out that over the past 20 years, less than 0.001 percent of the oil produced from the Outer Continental Shelf has been spilled from production facilities. There has not been a spill larger than 1,000 barrels from oil and gas platforms on the Outer Continental Shelf since 1980; in fact, natural seeps introduce approximately 100 times more oil into U.S. marine waters than do spills from offshore development and production activities. Today industry, the Department of the Interior's Minerals Management Service, and the U.S. Coast Guard are working in partnership to raise the bar for environmental and safety performance even higher. [65]

Further, the offshore oil and gas industry, including the support services sector, provides Americans with approximately 85,000 well-paying jobs, a number that is likely to more than double in the next two decades.[66] Oil production in the Gulf of Mexico, where there is a high level of industry interest and activity in waters as deep as 8,000 to 10,000 feet, is expected to double by the year 2002. Revenues from oil and gas development on the Outer Continental Shelf generate an average of $3 billion to $4 billion a year in federal receipts and help fund the Land and Water Conservation Fund and the National Historic Preservation Fund.[67]

In Russia, foreign energy companies believe they not only comply with Russian environmental regulations, but go even further by introducing even cleaner Western technologies and practices into Russian exploration. Exxon Neftgas Limited (ENL), operator of the large Sakhalin-1 Project, submitted a detailed, nine-volume Environmental Impact Assessment (EIA) of Phase 1 of the project, and was granted a positive conclusion by the State Ecological Expert Review of the Ministry of Natural Resources.[68] The Review Panel, comprised of 26 Russian experts from scientific, technical and environmental agencies of the Russian Federation, considered public input on the project, obtained from public consultation in all project areas, and received two public environmental expert reviews conducted by independent public organizations.[69] The Review Panel concluded that the scope and content of the documentation met the requirements of appropriate Russian laws and regulations, and demonstrated the environmental feasibility of the planned activities.[70]

As evidence of greater environmental commitment, the Sakhalin-1 Project offers examples of its positive record, including the following:[71]

- Use of world-class Extended Reach Drilling technology that allows wells located onshore to be drilled beneath the seafloor to oil and gas targets more than five miles offshore. This eliminates the need for some offshore structures, pipelines and associated activities;
- Drilling muds, cuttings and produced water from all Sakhalin-1 drilling platforms and drilling sites will be re-injected into the geological formations;
- Since 1994, the Sakhalin-1 Project has spent over $US 18 million on environmental field studies, both onshore and offshore;
- Since 1997, the Sakhalin-1 Consortium has spent about $8.5 million dollars on studies of the grey whales and measures to protect them. This research has expanded the knowledge base

about the species, and involved prominent Russian and Western whale scientists;

- The 2001 Odoptu seismic program, where the Project implemented the most extensive protection measures ever undertaken by industry. This included a 4-5 km protection zone between the seismic vessel and the grey whales and shutting down operations if the whales were present within this protection zone;
- To decrease the long-term presence of the Project along the coastline and reduce interference to near-shore grey whale feeding areas, the Consortium decided to remove permanent jetties from the Phase 1 project design at Chayvo;
- In 2004, the project initiated two programs to help protect eagles: Steller's Sea Eagle monitoring program and Artificial Nest and Perch program;
- The design and construction of the Sakhalin-1 pipelines will draw on the most advanced technology, and this technology has improved significantly over the last 25 years. Because the area where the pipelines will be laid onshore is not permafrost, pipelines will be buried;
- Pipeline design that accounts for a rupture strength in the case of seismic hazards with a return period of 1,000 years for onshore pipelines and 2,000 years for offshore pipelines;
- A leak detection system and an automatic shutdown system, including high accuracy flow and pressure meters at both intake and discharge ends; and
- Emergency shut-off (isolation) valves located every 30 km along the pipeline route.

Indeed, foreign energy companies have been accustomed to addressing environmental issues and often present evidence that the offshore operations can actually enhance the marine environment. For example, when Russian government negotiators sought compensation for prospective damages to the fish population in the waters surrounding Sakhalin Island, the energy companies offered studies to demonstrate that algae and other similar sea life that flourish around drilling rigs actually provide additional food for the fish population.

IV. THE INVESTOR'S SIDE: RISKS AND THE "HOSTAGE EFFECT"

Both the investor and the host government have incentives to pursue development of oil and gas reserves. But both face significant legal, economic and environmental risks. Balancing these incentives with these risks is the key challenge facing the investor and host government for the development of these projects. The goal of host governments is to first attract capital in order to develop the oil and infrastructure and the domestic economy as a whole. Once the investment is in place, many host countries turn more attention to both economic growth and protection of the environment. The goal of energy companies is to achieve a reasonable rate of return, long-term rights to produce and export oil and gas, fiscal and contractual stability for the life of the contract, and management of the operations on good commercial and environmental terms.[72]

There are special characteristics of an energy investment. First, the contract duration and the exposure to risk are long. Second, the amount of capital is enormous. And third, the types of risk are varied, from commercial to political to environmental. There are risks to property rights and land titles, risks of expropriation, risks of tort liability, risks of depletion and extraction, and currency risks. And one of the biggest problems facing energy companies investing abroad is the risk of unpredictable regulation.

Once an investment proves successful, a host country sometimes tries to change the terms of an investment relationship by imposing new taxes or environmental regulations. When profits are large, domestic public pressure and government priorities can cause a host country to push for a greater share of the investment returns than was provided in the original contract. Among the many techniques used to increase "government take" are the initiation of such taxes as environmental taxes to compensate for alleged extraction damage, taxes to help train the national workforce, taxes to improve public health and safety, or taxes to provide compensation to communities near the project site. Sometimes an environmental law that previously existed but has never been enforced is applied.

Once a project proves profitable, the energy company is subject to a "hostage effect," where the host government sees the deep pockets of Western energy firms and tries to extract more from them.[73] Because the energy firm has now sunk considerable fixed costs into the project, and also sees the proven revenue stream, abandoning the project is more difficult. Bargaining power shifts to the host government, and foreign investors are left to adjust to new circumstances.

V. THE HOST COUNTY'S SIDE: EFFICIENT ENVIRONMENTAL REGULATION

A host government is faced with the twin desires of attracting investment and growth while at the same time protecting its environment. At least three approaches are available for environmental regulation, including: (1) public law; (2) multilateral institution involvement; and (3) private contractual regulation.

a. Public Law

Host governments apply a wide range of regulatory regimes to investors in energy projects. At one end is the United States, which imposes detailed and strict environmental regulations on offshore oil and gas projects. Deborah Espinosa describes countries that are at the other extreme, such as Thailand.[74] Thailand's lax regulations provide that "the concessionaire shall take appropriate measures in accordance with good petroleum industry practice to prevent pollution in any place by oil, mud, or any other substance."[75] The wide array of regulatory regimes is related to the level of economic development in a country and its willingness to push for environmental concerns at the expense of attracting investment.

b. Multilateral Institutions

A second participant in environmental regulation is the multilateral institution, such as the World Bank. Multilateral development banks and other institutions often condition funding of energy projects on the satisfaction of various environmental criteria. This influence often results in significant improvements to a host country's environmental regulatory regime.

The World Bank has a group of environmental experts who assess the environmental consequences of all approved projects. For example, the World Bank is financing an oil project in Cameroon. The objective of the project is "to ensure a national capacity, in regards to environmental, and monitoring practices for the implementation of the Petroleum Development and Pipeline Project, which involves construction, and operation of an oil export pipeline, in the border with Chad."[76] The terms of the project include provisions to establish a regulatory framework, strengthen field inspections, train for spills and other accidents, and monitor the environment.[77]

Another example is in India, where the Bank has financed projects for the Bombay High Offshore Development Program.[78] The project involved development of the Bombay high oil field and included the drilling of development wells, the construction and installation of wells, processing and living quarter platforms, the installation of sub-sea flow lines, and engineering and consulting services for monitoring reservoir performance. The financing involved an environmental impact review.[79]

Another function of multilateral institutions involves project insurance. In particular, the Multilateral Investment Guarantee Agency (MIGA) insures investors against nationalizations and expropriations of foreign countries. Expropriation can include "creeping expropriation," the act of a host government gradually introducing taxes (such as new environmental taxes) and other policies that deprive the investor of his investment property. As an example, in 1997, MIGA provided $10 million in reinsurance to an investor for a portion of its coverage of an investment in the Algerian Rhourde el Khrouf oil field. The project is expected to continue production through 2004. The investment provides exploration services, production equipment for petroleum extraction in existing wells, and overall technical and managerial assistance.[80]

c. Private Contractual Regulation

A final approach toward environmental regulation of energy investments is to subject it to contract law. Described in great detail by Michelle Flores, contractual agreements between host countries and investing energy companies can provide their own framework for addressing environmental concerns:[81]

> Project agreements are thought to provide a simple administrative framework for allocating environmental responsibility. Parties usually include arbitration and choice of law language. The parties can arbitrate any violation of the agreement and enforce it like any other contract without involving the developing country's separate regulatory regime. Project agreements are solely controlled by private law principles (e.g., contract law).[82]

To address the "hostage effect," private contracts can include "stabilizations clauses." Such clauses try to restrict the application of unexpected environmental regulations and limit a host country to those laws already being enforced, and those provisions expressed in the contract. Or as an alternative, such clauses can leave open the possibility of new regulation, but require compensation to the investing party – a type of "renegotiation clause."[83]

VI. RUSSIA'S APPROACH

a. Background to Russian Environmental Law

Cymbre Van Fossen of the University of Wisconsin provides a useful history of Soviet environmental problems and regulation, and argues that post-Soviet Russia is only now coming to grips with the tension between economic development, popular belief in the infinite abundance of the "Rodina" or Motherland, and the real costs of neglecting the Russian environment. Although a love of their beautiful landscapes and mountains was a strong force in Russian culture, Van Fossen argues that the enormous size and natural wealth of the Soviet Union resulted in an "attitude of complacency towards resource depletion and pollution." Economic development under Stalin brought centralization of agrarian production and rapid industrialization. As many authors have described, the environmental record of the Soviet Union was abysmal.

The Soviet Union actually had one of the most advanced systems of environmental legislation in the world. By 1985, at least 670 separate environmental enactments were included in Soviet legislation. Enforcement of these legal provisions was, however, not a priority. One of the best descriptions of the Soviet environmental protection system comes from Deborah Espinosa, who writes:

> Imagine the United States without the Environmental Protection Agency, the Natural Resources Defense Council, and Greenpeace. Envision a chemical company drafting the Clean Water Act. That picture is in effect what was once the environmental policy of the Soviet Union. For approximately 70 years, the government of the USSR acted simultaneously as legislator, enforcer, violator, and adjudicator. Producers were

rewarded for their output regardless of the economic or social costs. That policy, combined with a prohibition on public participation, left the lands of the former Soviet Union poisoned, her resources depleted, and her citizens ill.[84]

b. Environmental Impact Assessments

Energy companies investing in Russia first encounter environmental protection law when bidding for contracts. Both Article 32 and Article 36 of the Law "On Environmental Protection" of January 19, 1991 (as amended through January 10, 2002)[85] and Article 18 of the Law "On Ecological Expert Review" of November 23, 1995,[86] require that before financing or implementation of projects that may cause harmful environmental effects may proceed, investors must first provide an expert commission with an environmental impact assessment (EIA).[87] The expert commission then either approves or rejects the EIA and forwards its conclusions to the Ministry of Natural Resources for confirmation.

The State Commission for Environmental Protection creates a commission of experts, which are comprised of staff of the Ministry of Natural Resources and outside experts.[88] The commission is given up to four months to evaluate the proposed project and make a decision. If less than two-thirds vote to approve the project, additional experts are asked to review the plan for an additional two months. If the evaluation is again negative, the project is rejected, but documents may still be submitted to address the reasons for rejection.[89]

c. The Law on Production Sharing Agreements

Russia's Law on Production Sharing Agreements (PSAs) imposed further environmental duties on investors in Russian energy.[90] While recent changes to Russian law will make it unlikely that further PSA contracts will be signed in the Russian energy sector, examination of the PSA framework is useful in understanding the development of Russian environmental law.

A production sharing agreement, or PSA, is a contract between a State and a private, usually foreign, investor, to exploit the State's natural resources and to divide production in contractually agreed proportions. First implemented by Indonesia in the 1970s, PSAs quickly became a favored vehicle for investment in developing countries with unstable or unclear legal regimes.[91] Believing that the 1991 Law on Foreign Investments and the 1992 Subsoil Law provided inadequate incentives to foreign investors, in 1993, President Yeltsin directed the Duma to draft legislation to permit the use of production sharing agreements in Russia. Two years later, in December 1995, the Russian Duma passed the controversial 1995 PSA Law.[92]

The PSA Law imposes environmental obligations on petroleum investors by requiring certain provisions to be included in every agreement. Article 7 of the PSA Law requires an investor to "take measures to prevent harmful impact of operations on the natural environment, as well as to remedy the consequences of such impact."[93] The law also required investors: (1) to "buy insurance against liability for damage caused in the event of accidents entailing harmful impact upon the natural environment;" and (2) to "remove all facilities, installations, and other assets upon completion of the agreement operations as well as clean up the territory on which the agreement operations were conducted."[94]

While the PSA Law imposes the environmental regulations discussed previously, it has been vague as to implementing regulations.[95] Another drawback is that while the law contains special legal stabilization clauses to protect investors from adverse changes in the law, the PSA Law expressly excluded from the scope of such stabilization any changes in health, safety, and environmental rules.[96] Such an omission raises the risk to investors of the "hostage-effect" discussed previously.

While the PSA Law was implemented in the early 1990s to facilitate investment in oil and gas, amendments to the law passed in June 2003 made it a less attractive vehicle for investors. In particular, an oil or gas field may become available for PSA development only after: (a) an auction for the development of such field on the standard license and tax terms has been held and has been declared void due to the lack of bidders; (b) the field complies with certain additional conditions introduced in the PSA Law (e.g., the fields are located in hard-to-reach areas lacking transportation and other infrastructure); and (c) this field has been included in the list of fields eligible for PSA development.[97] A separate auction will be held in order to select an investor with whom a PSA will be concluded in relation to any field from the list of eligible fields. The tender route for conclusion of PSAs was excluded.[98] These provisions make it unlikely that any more PSAs will be signed.

The only PSAs with foreign investment were signed before the enactment of the PSA Law and enjoy special protection under the "grandfather clause" of Article 2.7. Depending on the text of the individual PSA, that clause might be used to extend legal stabilization to cover changes in environmental law, thereby doing away with the exception for "health, safety and environmental regulation" in the stabilization provision of the PSA Law itself. This issue has not yet been presented in arbitration or in Russian courts so that the scope of the protection of the "grandfather clause" remains unclear.

d. The Pollution Charge System

Once a project is underway, it is subject to the environmental protection provisions of Russia's so-called "pollution charge system." Article 16 of the 2002 Environmental Protection Law allows companies to discharge hazardous substances subject to the periodic payment of a fee based on the type and amount of the pollutant.[99] The calculation and periodic adjustment of the fees is regulated by the Ministry of Natural Resources, which provides regional authorities with the ability to adjust, decrease, or eliminate the fees.[100]

Government Decree No. 632 of August 28, 1992, as amended in December 1997 and June 2001, describes the system for determining the fees. The system establishes a "Maximum Permissible Discharge" (MPD) level. For discharges below the MPD, the polluting companies pay a fixed fee based on roubles per metric ton. For discharges above the MPD, the fee is raised to five times this base level fine. The charges take into account ecological factors including "natural and climatic characteristics of certain territories, and the importance of natural features and socio-cultural objects" of the land.[101] There are over 210 identified pollutants, and the fees vary according to the ecological sensitivity of the body of water or the current air quality of the region in which discharge is contemplated and by adherence (or failure to adhere) to certain "maximum permissible concentrations."[102]

While the charge system acts as a form of environmental protection, it is also a significant revenue source for the government. Originally, the law provided that 90 percent of the fees charged was transferred to special accounts held by non-budgetary ecological government funds. The other 10 percent of the pollution fees were transferred to Russia's federal budget.[103] Of the 90 percent, 60 percent was spent locally, 30 percent regionally and the remaining 10 percent on federal environmental programs.[104] In 2000, the Russian Prosecutor General's office initiated an investigation into mismanagement of transferred funds. In October 2001, the Russian government disbanded the Federal Ecological Fund and all non-budgetary ecological funds. In a matter of weeks, Russia's Tax Ministry issued a letter explaining the new payment scheme: from that point on, polluters would transfer 19 percent of all fees due to the federal budget, and the remaining 81 percent to regional budgets.[105] Beginning in 2004, the federal budget claims 20 percent of such fees.[106]

e. Polluter Liability

An energy company can face legal liability both for pollution violations and for failure to obtain required permits. There are three types of liability the company may face, including: (1) civil liability; (2) administrative liability; and (3) criminal liability.[107]

Civil liability is usually found in situations where the defendant is at fault, either from negligence or deliberate acts. Strict liability is imposed for damage caused by "hazardous activities," one of which, according to Russian law, is the conduct of oil and gas exploration and production activities.[108] Unless a defendant can show the damage was caused by an act of God, civil liability will usually require a defendant to pay damages for injury caused to a person or property. These damages are normally actual damages plus lost profits and can include an injunction from continuing a project.[109]

Second, companies are subject to a wide range of administrative liability for any violation of environmental laws. Environmental inspectors are authorized to impose administrative penalties without approval or permission of the Ministry of Natural Resources. The penalties include fines, orders to terminate or suspend polluting activities, and orders to banks to discontinue financing of polluting projects.[110] One example of an administrative violation is failure to conduct proper environmental impact assessments.

Finally, individual members of the management of a company can be found criminally liable for serious violations that "cause significant harm to specific individuals or to the public at large."[111] Only courts may impose such liability on individuals, and the violation must be expressly described by an article of the Russian Criminal Code. Except for negligence that leads to a death, criminal liability is imposed only for intentional acts.[112]

CONCLUSION

The goal of this survey paper was to summarize the issue of environmental regulation in the Russian Federation, particularly that of offshore oil and gas. It is clear that there are major risks to oil and gas exploration, but the economic gains for Russia are also very great. The balance Russia chooses between economics and the environment will surely remain a topic for discussion in the coming years as more investment goes into Sakhalin Island and other regions.

Notes

[1] Economist Intelligence Unit (EIU), "Russia: Energy and Electricity Background," January 19, 2004. Available at Internet address: <www.eiu.com>.

[2] U.S. Department of Energy, Energy Information Agency, "Country Analysis Briefs: Russia," September 2003. Available at Internet address: <www.eia.doe.gov>.

[3] *Ibid.*

[4] Michelle Flores, "A Practical Approach to Allocating Environmental Liability and Stabilizing Foreign Investments in the Energy Sectors of Developing Countries," Colorado Journal of International Law and Policy, Winter 2001, p. 144.

[5] U.S. Department of Energy, Energy Information Agency, "Country Analysis Briefs: Russia," September 2003. Available at Internet address: <www.eia.doe.gov>.

[6] Deborah Espinosa, "Environmental Regulation of Russia's Offshore Oil and Gas Industry and its Implications for the International Petroleum Market," Pacific Rim Law and Policy Journal, July 1997, p. 651.

[7] *Ibid.*

[8] U.S. Department of Commerce, BASINS. Andrei Asimov, BASINS Representative in Khabarovsk, "Khabarovsk Krai Companies and Sakhalin Oil and Gas Projects," June 2003.

[9] U.S. Department of Energy, Energy Information Agency, "Country Analysis Briefs: Russia," September 2003. Available at Internet address: <www.eia.doe.gov>.

[10] Espinosa, supra at note 8, p. 652.

[11] U.S. Department of Energy, Energy Information Agency, "Country Analysis Briefs: Russia," September 2003. Available at Internet address: <www.eia.doe.gov>.

[12] U.S. Department of Energy, Energy Information Agency, "Country Analysis Briefs: Russia," September 2003. Available at Internet address: <www.eia.doe.gov>.

[13] U.S. Department of Commerce, BISNIS. Andrei Vasenyov, BISNIS Representative in Khabarovsk, "Khabarovsk Krai Companies and Sakhalin Oil and Gas Projects," June 2003, p. 2.

[14] Economist Intelligence Unit (EIU), "Government annuls tender for oil fields," February 2, 2004. Available at Internet address: <www.eiu.com>.

[15] Economist Intelligence Unit (EIU), "Russia's Latest Oil and Gas Oasis," May 13, 2003. Available at Internet address: <www.eiu.com>.

[16] Economist Intelligence Unit (EIU), "Five Years On, Any Better Off?," August 18, 2003. Available at Internet address: <www.eiu.com>.

[17] Economist Intelligence Unit (EIU), "How Oil Dependent is Growth?" January 19, 2004. Available at Internet address: <www.eiu.com>.

[18] Arkady Ostrovsky, "FT Report Russia: Growth is Strongest in Nearly a Century," Financial Times, April 13, 2004.

[19] Economist Intelligence Unit (EIU), "European Union: Energy Provision," December 30, 2003. Available at Internet address: <www.eiu.com>.

[20] Economist Intelligence Unit (EIU), "Moscow and EU Near Deal on Russia's WTO Entry," April 30, 2004. Available at Internet address: <www.eiu.com>.

[21] Economist Intelligence Unit (EIU), "European Union: Energy Provision," December 30, 2003. Available at Internet address: <www.eiu.com>.

[22] Economist Intelligence Unit (EIU), "Japan: Energy Provision," March 10, 2004. Available at Internet address: <www.eiu.com>.

[23] Economist Intelligence Unit (EIU), "Russia's Latest Oil and Gas Oasis," May 13, 2003. Available at Internet address: <www.eiu.com>.

[24] *Ibid.*

[25] Asia Pulse, "China to become promising LNG market by 2020," April 26, 2004.

[26] Economist Intelligence Unit (EIU), "Russia's Latest Oil and Gas Oasis," May 13, 2003. Available at Internet address: <www.eiu.com>.

[27] World Bank, "Russian Economic Report: February 2004," The World Bank, Russia Country Department. February 2004, 5-6. Available at <www.worldbank.org.ru>.

[28] *Ibid.*

[29] *Ibid.*

[30] *Ibid.*

[31] Economist Intelligence Unit (EIU), "How Oil Dependent is Growth?" January 19, 2004. Available at Internet address: <www.eiu.com>.

[32] Irina Baranova, "Leaking Operations: Environmental Consequences of World Bank and ERBD Investment in the Russian Oil Sector," CEE Bankwatch Network, p. 54. Available at Internet address: <www.bankwatch.org>. Retrieved May 7, 2004.

[33] CEE Bankwatch Network, "How the EBRD can prevent Sakhalin II environmental and social catastrophe," NGOs Issue paper for the EBRD Annual Meeting, London 2004, p. 1.

[34] Irina Baranova, "Leaking Operations: Environmental Consequences of World Bank and ERBD Investment in the Russian Oil Sector," CEE Bankwatch Network. Available at Internet address: <www.bankwatch.org>, p. 54. Retrieved May 7, 2004.

[35] See M.V. Pereladov, O. Yu. Vilkova, and A.V. Gebruk, "Background for the Strategic Environmental Assessment (SEA): Offshore Oil and Gas Development, Fisheries, and Biodiversity," chapter in Sakhalin Offshore Oil: Environmental Concern, Scientific Press, 2003.

[36] Baranova, supra at note 36.

[37] *Ibid.*

[38] *Ibid.*

[39] CEE Bankwatch Network, "How the EBRD can prevent Sakhalin II environmental and social catastrophe," NGOs Issue paper for the EBRD Annual Meeting, London 2004.

[40] U.S. Department of the Interior, "Offshore Discharges From Oil and Gas Development Operations – FAQ," Minerals Management Service (MMS) of the U.S. Department of the Interior, www.gomr.mms.gov.

[41] *Ibid.*

[42] Jonathan Wills, "Drilling Waste Streams from Offshore Oil and Gas Installations Drilling Fluids - What They Are and What They Do," Offshore-environment.com, p. 1.

[43] *Ibid.*

[44] Jonathan Wills, "Minimizing Waste Discharges and Their Effects" Offshore-environment.com.

[45] David Gordon, "Oil Industry a Threat to Sakhalin Economy and Ecology," Pacific Currents On-Line, December 2001, p. 1.

[46] *Ibid.*

[47] *Ibid.*

[48] See Centre for Russian Studies, "Oil Leak in Russian Tundra: Chronology of Events," Available at Internet address <http://www.nupi.no>.

[49] Baranova, supra at note 36, p. 57.

[50] *Ibid.*

[51] Trade and Environment Database (TED), "TED Case Studies: Russia Oil Spill," American University. Available at <www.american.edu/TED/komi.htm>.

[52] The World Bank Group, "Bank Helps in Russian Oil Spill," Available at <http://web.worldbank.org>, retrieved May 4, 2004.

[53] Joint UNDP/World Bank Energy Sector Management Assistance Programme, "Russia Pipeline Oil Spill Study," April 2003, p. 4. Available at <http://wbln0018.worldbank.org>.

[54] *Ibid.*, p. 5.

[55] *Ibid.*

[56] *Ibid.*

[57] U.S. Department of the Interior, "Seismic Activity and Acoustics," Minerals Management Service (MMS) of the U.S. Department of the Interior, www.gomr.mms.gov.

[58] Australian Conservation Foundation, "WDCS briefing on seismic survey noise pollution," Whale and Dolphin Campaign Center, available at: <http://www.acfonline.org.au>.

[59] *Ibid.*

[60] David Gordon, "Oil Industry a Threat to Sakhalin Economy and Ecology," Pacific Currents On-Line, December 2001, p. 1.

[61] Julia LeMense Huff, "Using the Tools We Have: An Integrated Approach to Protect the Sea of Okhotsk," Pace Environmental Law Review, Spring 2003, p. 694.

[62] *Ibid*, p. 695.

[63] BBC Monitoring Former Soviet Union, "Russia: Oil rig developer delays pipeline construction to protect whales," April 30, 2004.

[64] ExxonMobil, Environmental Policy, <http://www.exxonmobil.co.uk/files/pa/uk/Environment_policy.pdf>.

[65] Paul L. Kelly, Deepwater Offshore Oil Development: Opportunities and Future Challenges, National Oceanic and Atmospheric Administration, National Ocean Service. See <http://www.oceanservice.noaa.gov>.

[66] *Ibid*.

[67] *Ibid*.

[68] Sakhalin-1 Project, State Ecological Expert Review of the Ministry of Natural Resources Has Granted a Positive Conclusion to Phase One of the Sakhalin-1 Project, see <http://www.sakhalin1.com>.

[69] *Ibid*.

[70] *Ibid*.

[71] Sakhalin-1 Project, Safety, Health and Environment, see <http://www.sakhalin1.com>.

[72] Flores, supra at note 6, p. 146.

[73] *Ibid*, p. 161.

[74] Espinosa, supra at note 8, p. 664.

[75] *Ibid.*

[76] World Bank, "Cameroon Petroleum Environment Capacity Enhancement (CAPECE)," available at <http://web.worldbank.org>. Retrieved May 7, 2004.

[77] *Ibid.*

[78] World Bank, "Bombay High Offshore Development Project," available at <http://web.worldbank.org>. Retrieved May 7, 2004.

[79] *Ibid.*

[80] Multilateral Investment Guarantee Agency (MIGA), "2003 Regional Activities," available at <http://www.miga.org>. Retrieved May 7, 2004.

[81] Flores, supra note 6, p. 154.

[82] *Ibid.*, p. 155.

[83] *Ibid.*

[84] Espinosa, supra note 8, p. 656.

[85] Law of the Russian Federation on the Protection of the Natural Environment, January 19, 1991. Updated: Law of the Russian Federation on the Protection of the Natural Environment, No. 7-FZ, January 10, 2002.

[86] Law on Ecological Expert Review, No. 174-FZ, November 23, 1995; with Amendments of April 15, 1998.

[87] Directive of the Ministry of Natural Resources "On Confirmation of the Methodological Recommendations for the Preparation of Materials to be Presented to the State Ecological Examination Commission," No. 575, July 9, 2003.

[88] Order of State Ecological Commission of the Russian Federation: On the Regulation of the State Ecological Expert Review Process, No. N-280, June 17, 1997. Also see Deborah Espinosa, "Environmental Regulation of Russia's Offshore Oil and Gas Industry and its Implications for the International Petroleum Market," Pacific Rim Law and Policy Journal, July 1997.

[89] *Ibid.*

[90] Law on Production Sharing Agreements, No. 225-FZ, December 30, 1995; with Amendments and Additions of January 7, 1999 and June 6, 2003.

[91] Vitaly Timokhov, "Recent Developments in the Russian Production-Sharing Agreement Law: Making Sure the Law Works," UCLA Journal of International Law and Foreign Affairs, Fall/Winter 2001-2002, p. 365.

[92] Arina Shulga, "Foreign Investment in Russia's Oil and Gas: Legal Framework and Lessons for the Future," University of Pennsylvania Journal of International Economic Law," Winter 2001.

[93] Article 7, "Terms and Conditions for Conducting Operations," Law on Production Sharing Agreements, No. 225-FZ, December 30, 1995; with Amendments and Additions of January 7, 1999 and June 6, 2003.

[94] *Ibid.*

[95] Espinosa, supra note 8, p. 675.

[96] Article 17, "Stability of Terms of the Agreement," Law on Production Sharing Agreements, No. 225-FZ, December 30, 1995; with Amendments and Additions of January 7, 1999 and June 6, 2003. Also see John F. Sheedy, "Environmental Liability Under the Russian PSA and Subsurface Licensing Regimes," Coudert Brothers LLP, article appeared in the Russian Petroleum Investor, June/July 2001.

[97] Article 2, "Stability of Terms of the Agreement," Law on Production Sharing Agreements, No. 225-FZ, December 30, 1995; with Amendments and Additions of January 7, 1999 and June 6, 2003.

[98] Economist Intelligence Unit, "Changes to Production Sharing Agreements," July 14, 2003. Available at <http://www.eiu.com>.

[99] The fees themselves have been calculated and are detailed in the Order of the Government of the Russian Federation, "On Payment Levels for Atmospheric Pollution by Stationary and Mobile Sources, Pollution of Above- and Below-Ground Bodies of Water, and Storage of Industrial and Consumer Waste," No. 344, June 12, 2003.

[100] See Methodological Instructions for Levying of Fees for the Pollution of the Natural Environment, Instituted by the Ministry of the Environment, January 26, 1993; with Amendments and Adjustments of February 15, 2000. The "Instructions" also grant the regional authorities the same power to reduce or eliminate pollution payments. Polluters may be freed from the payments when, for example, the polluting substances are the product of work done for the benefit of "the socio-cultural sphere" (§7.5.1) or the production of heating and energy for the public (§7.5.3).

[101] Paragraph 2, Regime for Determination and Calculation of Payments for Environmental Pollution, Storage of Hazardous Substances, and other Dangerous Actions, Confirming Order of the Government of the Russian Federation, No. 632, August 28, 1992; with Amendments of June 14, 2001.

[102] Order of the Government of the Russian Federation, "On Rates of Payment for Dispersal into the Atmosphere of Polluting Substances from Stationary and Mobile Sources, Dispersals of Same into Above- or Below-Ground Waters, and Storage of Industrial Waste," No. 344, June 12, 2003. Also see Sheedy, supra at note 90.

[103] See orders of Ecological Ministry, Ministry of Finance, and State Tax Service of the Russian Federation, Nos. 04-15/61-638, 19, and BG-6-02/10, March 3, 1993.

[104] *Ibid.*

[105] See Bojan Soc, "Russia's Challenging Environment," Oil and Gas Eurasia, available at <http://www.oilandgaseurasia.com>, p 4. Retrieved May 8, 2004.

[106] Order of the Government of the Russian Federation, "On Rates of Payment for Dispersal into the Atmosphere of Polluting Substances from Stationary and Mobile Sources, Dispersals of Same into Above- or Below-Ground Waters, and Storage of Industrial Waste," No. 344, June 12, 2003. Regime for Determination and Calculation of Payments for Environmental Pollution, Storage of Hazardous Substances, and other Dangerous Actions, Confirming Order of the Government of the Russian Federation, No. 632, August 28, 1992; with Amendments of June 14, 2001.

[107] Order of the Government of the Russian Federation, No. 344, see, supra, at note 100. Also see Sheedy, supra note 90.

[108] Art. 1079, Civil Code of the Russian Federation.

[109] Sheedy, supra note 90.

[110] *Ibid.*

[111] Order of the Government of the Russian Federation, No. 344, see, supra, at note 100.

[112] Criminal Code of the Russian Federation, No. 63-FZ, June 13, 1996, §§109, 111, 250-251.

Energy Policy and International Royalty: A Dormant Servitude Relevant for Offshore Development*

Aldo Chircop†

INTRODUCTION

In recent years offshore oil and gas drilling technology has developed to enable exploratory activities on the slope and rise of the continental margin outside the 200 nautical mile (M) limit. It is likely that technology will continue to develop rapidly so that any commercial finds may be brought into production within the foreseeable future. There are many areas on the continental margin outside the 200M limit in all oceans that are thought to be promising for various non-living resources, especially hydrocarbons.[1] Some believe methane hydrates on the continental margin are likely to be among the largest natural gas reserves in the world.

State parties to the United Nations Convention on the Law of the Sea, 1982 (UNCLOS), are entitled to exercise sovereign rights for the purpose of exploring and exploiting non-living natural resources and sedentary species up to 200M of continental shelf or, where the continental margin extends beyond 200M, up to the limits of the continental margin according to criteria set out in that instrument.[2] There may be as many as 50 or more states entitled to exercise sovereign rights over an extended continental shelf; in this article these are referred to as broad margin States.[3] Most of the eight Arctic Council states are likely to be entitled to extended continental shelf rights. At the time this article was written, UNCLOS had been ratified by Canada, Denmark (Greenland), Finland, Iceland, Norway, the Russian Federation and Sweden.[4] Denmark, whose Arctic presence is

* This paper is adapted from an earlier work of the author entitled "Operationalizing Article 82 of the LOS Convention: A New Role for the International Seabed Authority?" 18 *Ocean Yearbook* (2004) (by The University of Chicago. All rights reserved). The permission of the University of Chicago Press to use the earlier work is gratefully acknowledged. The author is also grateful to David Moulder, Librarian and John Liljedahl, Lecturer, both at the World Maritime University for assistance with specific queries.

† Canadian Chair in Marine Environment Protection, World Maritime University, Malmö, Sweden.

by virtue of Greenland, and the United States are expected to ratify in the near future.[5] Arctic broad margin States Canada, Denmark, Norway, the Russian Federation and the United States have undertaken or will undertake scientific research that will assist eventual submissions to the Commission on the Limits of the Continental Shelf. Of these, it is only the Russian Federation that has so far proceeded with an actual submission to the Commission. Without waiting for formal determination of seaward limits, licences for exploration or discovery activities are already being issued on the Canadian Atlantic extended continental shelf. Exploration licenses have also been issued for the U.S. Gulf of Mexico extended continental shelf. The Arctic is a much harsher environment and it is likely that it will be a considerably longer period before similar licenses will be issued in Arctic extended continental shelf regions. Production on the extended continental shelf has not yet taken place, but discoveries should be expected and production could well occur within the decade, at least on the northwest Atlantic continental margin. When such production occurs, Article 82 (Payments and Contributions with respect to the Exploitation of the Continental Shelf beyond 200 Nautical Miles), an obscure and innovative provision in UNCLOS, will commence to operate.[6] It prescribes a novel duty for broad margin States to make payments or contributions in kind on the production from non-living resource exploitation. That revenue or in-kind contribution will be made available to UNCLOS State parties, especially developing countries, through the International Seabed Authority (ISA).

There are several legal and policy issues that can be expected to arise as a result, in particular the interpretation of Article 82, the role of the ISA, and domestic implementation. Different interpretations may potentially have weight in determining commercial feasibility. Although the role of the ISA appears to be defined in narrow terms, the criteria for the distribution of benefits remain to be defined. At this time it is fair to state that broad margin States are as unprepared for the implementation of Article 82 as the ISA. Article 82 contains many ambiguities, and the Convention and its diplomatic history provide little guidance on the operationalization of this provision. This paper identifies key issues of interpretation underlying Article 82 and concludes that the implementation of this article will necessitate close cooperation between the ISA and pioneering broad margin States.

ARTICLE 82

Article 82 states the following:

> 1. The coastal State shall make payments or contributions in kind in respect of the exploitation of the non-living resources of the continental shelf beyond 200 nautical miles from the baselines from which the breadth of the territorial sea is measured.
>
> 2. The payments and contributions shall be made annually with respect to all production at a site after the first five years of production at that site. For the sixth year, the rate of payment or contribution shall be one per cent of the value or volume of production at the site. The rate shall increase by one per cent for each subsequent year until the twelfth year and shall remain at seven per cent thereafter. Production does not include resources used in connection with exploitation.
>
> 3. A developing state which is a net importer of a mineral resource produced from its continental shelf is exempt from making such payments or contributions in respect of that mineral resource.
>
> 4. The payments or contributions shall be made through the ISA, which shall distribute them to States Parties to this Convention, on the basis of equitable sharing criteria, taking into account the interests and needs of developing states, particularly the least developed and the land-locked amongst them.[7]

During negotiations at the Third United Nations Conference on the Law of the Sea (UNCLOS III) there was a close trade-off relationship between Article 76 (Definition of the Continental Shelf) and Article 82.[8] For most states the normal entitlement to exercise sovereign rights over the legally defined continental shelf covers an area up to a limit of 200M from the baselines from which the territorial sea is measured. Broad margin States received more. The special entitlement of broad margin States in relation to the extended continental shelf came about at the expense of the International Seabed Area (the Area) and common heritage of mankind. When he launched the revolutionary concept of the common heritage of mankind in 1967, Maltese Ambassador Arvid Pardo proposed that the Area be developed in the interests and for the benefit of all mankind, but with preferential consideration for developing countries.[9] He also proposed that states that were to benefit from the new maritime zones should contribute a percentage of the revenues generated from the development of the living and non-living resources within their jurisdiction.[10] He suggested that there be equitable sharing of the benefits generated from development in the Area and a percentage of the revenues derived from national activities. This thinking carried through the Seabed Committee and UNCLOS III, except that it eventually focused on the extended continental shelf only. Several states, in particular the land-locked and geographically disadvantaged, resisted the seaward expansion of maritime zones, which they perceived as an encroachment on the Area.[11] Proposals to subject the EEZ and continental shelf within the 200M limit to the Article 82 regime were eventually abandoned. The "package deal" procedure of UNCLOS III and the effort at producing consensus eventually led to a compromise. Coastal States, and broad margin States in particular, agreed to an application of the Article 82 regime to the extended continental shelf in return for agreement on the new seaward limits of the continental shelf in Article 76. As a result, the continental shelf windfall benefit for broad margin States was accompanied by the contingent duty to make payments or contributions according to a predetermined formula based on production within national jurisdiction, but beyond 200M.

NATURE OF THE OBLIGATION

Article 82 is a negotiated international obligation. UNCLOS III negotiators stopped short of calling the obligation an international "tax."[12] Tax is a well-defined concept in a domestic context with connotations that may not be relevant or helpful for application in a contemporary international setting, such as the exercise of the sovereign's right to levy monies so that government may be supported and services provided. An essential characteristic of tax is that "it is not a voluntary payment or donation, but an enforced contribution, exacted pursuant to legislative authority."[13] Also, the power to tax connotes a power to assess (i.e., assessments of tax due), and the ISA has not been duly empowered by UNCLOS to make assessments in relation to Article 82. The broad margin State does not make the payment or contribution to the ISA, nor does it enjoy any special service entitlement as a result of the payment or contribution. Also, considering that the payment or contribution relates to activities within national jurisdiction, the utilization of tax to conceptualize the new international obligation would have established a precedent that State parties were not ready for at the time. Taxation of activities undertaken as sovereign rights is in itself a *sovereign* power of the state, and thus it is unlikely that negotiating states would have agreed to the creation of a taxation function in a global international organization. Even in relation to activities in the Area, there is no utilization of domestic law fiscal concepts connoting the exercise of sovereign functions, and instead reference is made to "payment," a generic concept connoting a financial obligation and no more.[14] The references to "payment" and "contribution" do not evoke ideas of sovereign prerogatives, but simply convey the content of an international obligation.

Rather than an international tax, it is probably more appropriate to treat the Article 82 obligation as a type of "international servitude," although in practice this term is rarely used "because of its emotive suggestion of dominance and servience."[15] Servitude is a domestic civil law concept connoting a burden on real property.[16] At international law, Article 82 is akin to an economic servitude, tied to a specific type of territory (i.e., the extended continental shelf), creating a restriction on it and applying for the benefit of other UNCLOS State parties.[17]

From oil and gas industry and regulatory perspectives, Article 82 is likely to be perceived as a royalty. In general, a royalty constitutes "compensation for the use of property . . . (e.g., natural resources) . . . expressed as a percentage of receipts from using the property as an account per unit produced," and more specifically in "mining and oil operations, a share of the product or profit paid to the owner of the property."[18] In offshore development, where the "property owner" is the state rather than a private individual, a royalty is a payment to the crown in consideration for the licensed utilization of a natural resource. The oil and gas royalty regime is a complex legal and management structure that is generally legislated, but whose implementation may also be negotiated on a per-licence basis. Much depends on the anticipated costs of the exploration, development and anticipated production from a particular field. The technological and transportation costs for deep-sea development are high. Given the risks and associated costs, royalty expectations may well determine whether a particular discovery is in fact a commercial find at all.

Another similarity to the domestic royalty is the element of compensation, in the case of Article 82 presumably for International Seabed Area "lost" to the broad margin State as a result of Article 76. However, there is also a significant difference: the payment and contribution in kind are due not to the licensing state, and not even to the ISA, but to UNCLOS State parties. In financing or otherwise fulfilling its obligation, one option that the broad margin State can be expected to consider is its domestic royalty regime.

The text of Article 82 may also provide for another and more flexible interpretation of how the obligation may be discharged, which might point to another means of fulfilment of the obligation. Although the payment or contribution would be based on the value or volume of production, there is nothing to suggest that this obligation is not equally discharged if the broad margin State elects to fulfil the obligation, to the same material extent, through some other manner, for instance the provision of Official Development Assistance (ODA). The reference to "in kind" can be read as inclusive of various available options for the fulfilment of the obligation. However, in this scenario it would be important that the performance of the legal obligation does not amount to the conversion of an existing prior responsibility, but rather must be a direct expression of the obligation in Article 82. In performing the conventional obligation in good faith, the broad margin State would need to demonstrate that it is responding to the obligation consistently with what was intended by the UNCLOS III negotiators.

ELEMENTS OF ARTICLE 82

Basic Principle: Duty to Make Payments or Contributions in Kind

The choice between performing the obligation through financial payment or contribution in kind rests with the broad margin State. It is not clear who is responsible for determining the precise percentage of value or volume that may be due. It could be, for example, the broad margin State, the producer in the broad margin State, or the ISA. The producer and broad margin State are ostensibly better informed than the ISA as to the volume or value to be taken into account. The ISA has not been mandated with an assessment power, but it is responsible for receiving the payment or contribution. Accordingly, it is reasonable to suggest that the ISA (and possibly other UNCLOS State parties) should be informed of the basis for the calculation of the amount or payment that is due.[19]

In some states there is likely to be an issue as to who is ultimately fiscally responsible for the payment or contribution in kind. While the obligation in Article 82 is international law, its domestic implementation may raise concerns as to where the ultimate responsibility lies. In a federal State, the answer to this question may be even more complex, especially where a state or a provincial government may have entitlements or expectations to a share of offshore revenue, together with the federal government. Because of its constitutional history, this is a potential issue for Canada because the federal and provincial governments have entered into an agreement on the levying of royalties and sharing of revenues.[20] The issue that could arise is whether the obligation at international law is one that ought to be shared internally by those levels of government that benefit from offshore development and industry. It would be unrealistic not to anticipate that the Article 82 obligation would not be transferred, at least to some measure, to the offshore producers. Accordingly, in some countries it is going to be important for legislators and offshore licensing agencies to anticipate how the cost of this international obligation is going to be met. It may well be that federal, state or provincial governments, and exploration licensees may have to reconsider the costs associated with offshore development and production on the extended continental shelf.

Although not yet a party to UNCLOS, the United States appears to be the first broad margin State to consider how the royalty might be levied. In recent Lease Stipulations for the Gulf of Mexico issued by the Minerals Management Service (MMS), lessees are advised that contingent royalty payment lease provisions would apply if the United States were to become a party to UNCLOS prior to or during the life of a lease. [21] The Lease Stipulations are clear that "[a]ny such payments will be made by the U.S. government and not the lessee."[22] However, this appears to refer to the United States' obligation as a coastal State and does not *per se* indicate that the federal government will actually pay the royalty out of federal resources. Thus the Lease Stipulations provide for lessees to pay the "Convention-related royalty . . . so that the required Convention payments may be made by the U.S. government as provided under the Convention . . . ."[23] Interestingly, some interests in the U.S. oil and gas industry, while considering the obligation as a "modest" revenue-sharing provision, have the understanding that "this royalty should not result in any additional cost to industry."[24] Those same interests also expressed concern with the lack of detail on how the revenue-sharing will work, and the consequence is some uncertainty for industry.[25]

Article 82 foresees the making of payments or contributions on an annual basis. Consistently with this requirement, the Lease Stipulations require Convention-related royalty payments to the MMS after the expiration of a production lease year.[26] Presumably, the obligation is discharged if a representative domestic public authority makes the payment on behalf of the broad margin State. It may be possible for the broad margin State to indicate that the obligation to make the payment or contribution (especially where the contribution option is exercised) is to be discharged by the domestically licensed producer directly. In any case, it is likely that the value of the payment or volume of the contribution will vary from year to year in response to market and other conditions.

Applicable Resources

Early proposals in anticipation of and during UNCLOS III included all resources of the EEZ, but by the time the first negotiating text was produced it was clear that the duty would have a limited resource and spatial application.[27] Coastal States became increasingly reluctant to share resources or related benefits that were otherwise subject to the exclusive sovereign rights of exploration and exploitation. At the same time, those states not having a broad margin resisted an unconditional expansion of coastal State rights. Insofar as the EEZ was concerned (and continental shelf up to 200M), the idea of sharing was whittled down to the surplus of fisheries (but not sedentary species) in that zone.[28]

In the case of the continental shelf, the idea of sharing was eventually confined to the non-living resources of the extended continental shelf.[29] As seen earlier, the rationale was that the broad margin State was going to enjoy an additional benefit beyond that enjoyed by other coastal States, and in effect at the expense of the common heritage of mankind. Mankind, perhaps in the form of the other State parties to UNCLOS, was entitled to benefit from this encroachment on the area, not necessarily with reference to the size of the extended shelf falling within national jurisdiction, but rather in relation to the benefits of resource development. This was consistent with the idea that benefits to be derived from the utilization of the International Seabed Area and its resources were to be derived for the benefit of mankind as a whole, albeit with preferential consideration for some states and peoples.[30]

It can be expected that Article 82 will also apply to transboundary non-living resources straddling the following: (1) the EEZ and extended continental shelf, (2) neighbouring extended continental shelves, and (3) the extended continental shelf and International Seabed Area. Where (1) occurs, the producing state would need to determine what percentage of the production is proportionate to that part of the resource located on the extended shelf. This scenario could pose considerable difficulties for producers because it may well be that only part of the production would be subject to the payment or contribution, when in reality the resource would be developed as a unit. Difficulties could be exacerbated where the payment or contribution could be considered by industry as discriminatory because it would obviously not apply to offshore development within the 200M limit. In the case of (2), if the transboundary resource development is unitized, presumably the duty will apply to the overall production from the unitized field or joint development zone, and the neighbouring states would have to agree on their respective shares of payments or contributions just as much as they would normally agree on benefit-sharing. On the other hand, if the development is unilateral, then the payment or contribution is clearly the duty of the producing state only. In the case of (3), there could be two scenarios. The first is one where the broad margin State undertakes unilateral development of the transboundary resource, in which case it would have to make the payment or contribution in relation to production.[31] The second scenario is where the ISA, perhaps through a future-established enterprise or some other mechanism, enters into joint development with the broad margin State. In this case there would have to be agreement on what proportion of the resource subject to the payment or contribution is located on the extended continental shelf. Although not the direct concern of this article, a joint development zone involving the Area in part will need to satisfy the procedures for the exploration and exploitation in Part XI, although the ISA's jurisdiction would be limited to those activities specifically occurring in the Area.[32]

Basis of Assessment and Rising Scale of Payments or Contributions

Article 82 foresees a grace period of five years during which the producer is expected to recoup the development costs. On the sixth year the payment or contribution will commence to operate at the rate of one percent for that year, and will continue to increase by one percent per year until the ceiling of seven percent in the 12th year is reached. This last percentage will then continue to apply over the rest of the production period. This formula was probably based more on the need for a compromise than on the realities of cost recovery. There were alternative proposals on the negotiating table. Early proposals would have levied different percentages on the production depending on whether it occurred within or outside 200M. Others included contributions of up to five or 10 percent.[33]

The grace period may cause difficulties for producers in two ways. First, given the very expensive modern-day offshore development environment, it remains to be seen whether five years is sufficient to recoup development costs in their entirety. The answer to this question will depend on the field concerned, the difficulty of the geology, market value of the product and regulatory obligations. Second, it appears that the grace period is applicable for the recovery of the development costs only. Once the payment or contribution period commences, the assessment appears to be on the gross value, rather than on net value. This means that although costs are significantly lower at the production stage, there is no deduction of these costs before the royalty is determined. It seems that this was a conscious addition to this provision. A proposal in a Working Paper to apply the royalty on the net did not receive sufficient support.[34] The U.S. delegation considered this proposal but felt that there would be significant difficulties in dealing with the accounting processes needed to determine the net, especially considering the differences between economies.[35]

Eligible Contributing States

Not all broad margin States need make payments or contributions. During UNCLOS III there were differences of opinion as to whether all states should be subject to the obligation, but the compromise reached was that developing states that are net importers of the mineral resource produced on their continental shelves would be exempt.[36] Therefore, a developing state that exports the mineral resource concerned would be subject to the obligation. A producing but net importer state could be in a position to benefit from the payments and contributions made by other states. Whether such a state would be entitled to benefit *pari passu* with other developing states will depend on the equitable sharing criteria developed by the ISA.

Role of the ISA

Delegations participating in the Second Committee at UNCLOS III were divided over the role that an appropriate international organization would play in relation to Article 82, but there was eventual compromise in support of a role for the ISA.[37] UNCLOS provides little guidance on the actual role of the ISA. In part this is due to the nature of the duty articulated in this provision, namely a responsibility for broad margin States, and a collateral role to be played by the ISA in fulfilling the objectives of the provision. Even so, the ISA's role and its relationship to the broad margin State need to be clarified.

The Preparatory Commission (PrepCom) for the ISA and International Tribunal for the Law of the Sea (Tribunal) established under Annex 1 of UNCLOS did not consider this provision during its 10-year period of work.[38] The mandate of PrepCom consisted of tasks that would enable the ISA and the Tribunal to commence functioning when that instrument came into force. Its output consisted of various recommendations, rules, procedures, agreements and studies for (1) the Assembly and Council, (2) the Secretariat, and (3) States parties. Article 82 considerations did not constitute a specifically assigned task and PrepCom's work focused on organizational, structural, informational and regulation-drafting tasks that would enable the ISA to function without delay once the Convention came into force. Issues that would have required negotiation on behalf of the future ISA with specific states could only be addressed if specifically mandated by UNCLOS III.[39] The role to be played by the ISA will need to be inferred in part with reference to Article 82 itself and its negotiating history, in part with reference to other provisions in UNCLOS, and with reference to what may be implicitly inferred from the purposes of Article 82 and the overall functions of the ISA.

The Role of the ISA as Stated in Article 82

Paragraph 4 suggests both passive and active roles for the ISA. The passive role is with reference to the broad margin State in the sense that it is the recipient of payments or contributions made by the broad margin State, but there is no role allocated for the determination of their nature, amount, tradable currency and precise timing. The active role is with reference to what happens to the payments and contributions once made, but the ISA's role appears to be limited, as was pointed out recently in the

ISA Assembly.[40] The ISA is to distribute these on the basis of equitable sharing criteria mentioned earlier. Clearly the responsibility to establish the criteria is a significant task. But even here, the ISA would have to perform this task within the framework of general criteria already provided. Also, it is important to note that the provision does not suggest that any of the payments or contributions may be retained by the ISA to cover, for example, administrative costs. There is no mention that anything less than full payments or contributions (and thus not net proceeds) must be distributed to States parties.

As noted earlier, the broad margin State's payments or contributions are to be made *through* the ISA (note: not *to* the ISA) for the purpose of distribution to State parties. Clearly, the ISA is not the final destination of these payments or contributions. Pardo's original idea before and during the Seabed Committee was that the payment would be made to the International Ocean Space Institutions that would emerge from UNCLOS III. During the negotiation of this provision, there were proposals designating the ISA as a beneficiary of the payments or contributions, but they did not receive significant support to enable inclusion in the negotiating texts.[41]

There is thus a fettered role that can be played by the ISA, perhaps in the form of a trustee of funds or resources. First, the ISA is the institution specifically designated to receive the payments and contributions, which presumably will be held in trust and on a temporary basis.[42] Whether this function entails an implicit subsidiary function for the ISA to determine if and when the obligation of a broad margin State has been discharged is open to conjecture. Second, the ISA has the mandate to distribute the payments and contributions to States parties. Third, because distribution must occur on the basis of equitable sharing criteria, it is reasonable to presume that the ISA also has the authority to determine those criteria.

Powers and Functions of the ISA in Other Pertinent Provisions

UNCLOS foresees roles for both the Assembly and Council of the ISA. The Assembly is the supreme organ of the ISA and is empowered "to establish general policies in conformity with the relevant provisions of this Convention on any question or matter within the competence of the ISA."[43] The Council is the ISA's executive organ and has the power to establish specific policies on any matter within the ISA's competence, but within the more general policies of the Assembly.[44] This mandate is further augmented by other assigned powers, including, for Article 82 purposes, the power to

> . . . recommend to the Assembly rules, regulations and procedures on the equitable sharing of financial and other economic benefits derived from activities in the Area and the payments and contributions made pursuant to Article 82, taking into particular consideration the interests and needs of developing States and peoples who have not attained full independence or other self-governing status . . . .[45]

In turn, the Assembly will consider those recommendations and may approve them or "return them to the Council for reconsideration in the light of the views expressed by the Assembly."[46]

In principle, the procedure appears simple. However, it is unclear how differences would be resolved (other than by resorting to formal dispute settlement) between a broad margin State and the ISA if there were differing interpretations of the full responsibilities of such a state, including the payment or contribution due. The ISA would need to be informed of the nature and extent of eligible activities on extended shelves, but there is no explicit duty to inform on the part of the broad margin State. Again, there is no explicit mandate for the ISA to monitor activities (other than desktop exercises) on extended shelves or, even more far-reaching, to audit production to ascertain or corroborate the full extent of the payment or contribution.[47] In effect, the ISA would have to rely on the goodwill of the broad margin State, because it is possible that the broad margin State may not be compelled to disclose.[48] Admittedly, broad margin States have an obligation to "fulfil in good faith the obligations assumed" under UNCLOS, and to avoid exercising their rights in a manner that would constitute an abuse of rights.[49]

It is conceivable that although the Council has to recommend a framework to the Assembly for Article 82 purposes, there may have to be *ad hoc* arrangements between the ISA and an individual broad margin State. Such arrangements may need to address how the obligation will be performed (payment or contribution), the value or volume calculated, modalities of payments or contributions established, and how timing of payments is to be determined.

Possible Implicit Functions

Whether the ISA enjoys functions other than what is specifically mandated in UNCLOS could be a moot point. As a treaty creation, the ISA is required to function with the framework and boundaries of its empowering instrument. Article 157(2) contains the following:

> The powers and functions of the ISA shall be those expressly conferred upon it by this Convention. The ISA shall have such incidental powers, consistent with this Convention, as are implicit in and necessary for the exercise of those powers and functions with respect to activities in the Area.[50]

The incidental powers seem to be limited to activities in the Area. The receipt of payments or contributions in Article 82 may be seen as a function that is unrelated to activities in the Area and therefore the ISA may not enjoy incidental powers in this regard, other than what is specifically provided in UNCLOS.

A possible alternative view is that although the Article 82 obligation is established outside Part XI, the ISA's role in relation to this article is also spelled out in Part XI. The interpretation of the content of Article 82(4) and other relevant provisions of UNCLOS should be guided by the spirit behind these provisions. The ISA was given a role and will need flexibility to fulfil that responsibility pragmatically and effectively with due regard to the intent of Article 82, even though it may be argued that the ISA has no incidental powers in this regard.

Equitable Sharing: Eligible Beneficiary States

In principle, all States parties to UNCLOS are entitled to benefit from the payments and contributions made. The sharing process is expected to be governed by equitable sharing criteria adopted by the ISA. The only UNCLOS *a priori* criteria to govern equitable sharing are: (1) the taking into account of the interests and needs of developing states; (2) the particular consideration of the interests and needs of least developed and land-locked developing states; and (3) where Article 162(2) further mentions "the interests and needs of developing States and peoples who have not attained full independence or other self-governing status."

Insofar as (3) is concerned, UNCLOS seems to contain some minor inconsistency between Article 82(4) and Article 162(2) in the way preferential consideration is put forward.

It is not clear what "taking into account" means, but it does imply preferential consideration. The intent of "interests and needs" is also unclear. Is it conceivable that a hierarchy of needs may be established by the ISA as criteria to determine access or extent of access to the payments and contributions (e.g., states with basic livelihood needs, in comparison with states that wish to reduce their dependence on imported energy)? In theory, it is conceivable to foresee a tentative general ranking, in order of priority, as follows:

(1) least developed land-locked states (and presumably peoples who have not attained full independence or other self-governing status);
(2) other developing states (including other land-locked developing states); and
(3) other states (including newly industrialised, developed land-locked and developed states generally).

The ISA would need to do more than this. It would need to consider which states would be considered least developed and developing. It might have to consider socio-economic criteria and mineral resource dependence. Since the adoption of UNCLOS there are now also newly industrialised states, and it remains to be seen how these will be characterised for benefit-sharing purposes. Also, the negotiating history of UNCLOS III and the requirement to consider equitable criteria in an UNCLOS context suggest that the situation of developed land-locked states and geographically disadvantaged states may also have to be considered.

DISPUTE SETTLEMENT

UNCLOS is silent on the consequences of a broad margin State's refusal to make or unreasonably delay in making the required payment or contribution in kind. Although during the negotiation of Article 82 there were suggestions to consider the consequences of non-payment (including determination of when non-payment occurs), there was no consensus on a rule that would tighten the broad margin State's obligation.[51]

It is unclear what dispute settlement mechanism should apply to a dispute between a broad margin State and the ISA. State parties have access to the dispute settlement institutions set up by UNCLOS, but the access of the ISA to the same mechanisms is constrained: "The dispute settlement procedures specified in this Part [i.e., Part XV on dispute settlement] shall be open to entities other than State Parties only as specifically provided for in this Convention."[52] Clearly the ISA has access to the Tribunal's Seabed Disputes Chamber,[53] so long as the claim by the ISA is eligible subject matter. The jurisdiction of the Chamber is restricted to disputes involving activities in the International Seabed Area and with reference to specific categories.[54] A dispute on the interpretation or application of Article 82 does not concern activities in the Area. Even if the dispute concerns the Assembly's and Council's regulatory powers in relation to Article 82, these powers do not concern activities in the Area. In any case, even where the Chamber has jurisdiction, there is a significant limitation on the exercise of that jurisdiction. Where a dispute is concerned, the Chamber "shall not pronounce itself on the question of whether any rules, regulations and procedures of the Authority are in conformity with this Convention, nor declare invalid any such rules, regulations and procedures."[55] This is pertinent when the Council's and Assembly's powers to make rules, regulations and procedures in relation to Article 82 are concerned.[56]

There is an argument for considering an Article 82 dispute as one encompassed by Article 187 only if the dispute concerns a resource that straddles an extended continental shelf and the Area. However, where differences between the broad margin State and the ISA relate to the interpretation of Article 82, the dispute settlement route is not stated. Part XV is concerned with disputes between states as well as Part XI Section 5 disputes, including "such a dispute" involving an entity other than a State party.[57] Despite the intention of the negotiators of UNCLOS III to extend the application of Part XV Section 1 to disputes between a State party and the ISA,[58] the technical reference to "such a dispute" (i.e., qualifying the dispute to one relating to activities in the Area), seems to exclude Article 82 disputes.

Although the drafting of Article 285 might pose a technical hurdle for Article 82 disputes, it is essential that the broad margin State and the ISA are willing to resolve the dispute. Article 285 was negotiated with the spirit and intention of enabling State parties and the ISA to resolve their disputes any way they choose.[59] Further, the provisions on the Tribunal's competence anticipate the possibility of State parties and other entities appearing before it, either as provided in Part XI or "in any cases submitted pursuant to any other agreement conferring jurisdiction on the Tribunal which is accepted by all the parties."[60] This text is generic enough not to exclude the possibility that "any other agreement" could be a special agreement between a broad margin State and the ISA conferring jurisdiction on the Tribunal.

There is also the possibility of a resort to conciliation, but probably not arbitration under UNCLOS.[61]

CONCLUSION

As has been seen, UNCLOS provides relatively scare guidance on how Article 82 should be interpreted, let alone implemented. Broad margin States with an interest in this article and the International Seabed Authority have complementary roles to play in implementing this provision. Neither individual broad margin States, nor the ISA alone will be able to make this provision work without close consultation and agreement. Insofar as the broad margin State is concerned, such a state will need to integrate Article 82 responsibilities in its domestic royalty regime. It needs to anticipate how it will fulfil its "sharing" responsibility. This needs to be done sooner rather than later, because the time scales for offshore development tend to be over the long term, perhaps over decades. Industry needs to fully appreciate the extent of its fiscal obligations when investing in extended continental shelf activities. The fact that offshore fiscal regimes tend to be negotiated well ahead of actual production indicates some urgency for broad margin States that are already licensing offshore exploration activities on the extended shelf in order to avoid potential domestic discord.

The ISA, on the other hand, is potentially poised to play a more active role in channeling benefits to the developing countries than currently possible under the bulk of its mandate in Part XI. In order to do so, the ISA needs to start considering how it could perform its Article 82(4) responsibility by utilizing the powers granted to the Assembly and Council in Part XI. Council and Assembly members will need to interpret the explicit and implicit mandate conferred by UNCLOS to enable the ISA's performance of its responsibility, both with regard to paying or contributing broad margin States and in relation to beneficiary States parties. The criteria for sharing benefits need to be developed. At the same time, and in interpreting its mandate, the ISA will need to work closely with interested broad margin States to clarify mutual expectations and minimize uncertainty that might discourage offshore development.

Notes

[1] A more reserved view on this is reflected in a recent report of the ISA Secretary-General: ". . . according to current knowledge, the offshore resources in potential continental shelf areas beyond 200 nautical miles are sub-marginal to para-marginal, technological improvements in recovery efficiency and greater access to deep-water areas are already increasing the range of economically recoverable resources and there is considerable potential for exploitation of these resources in the future." Report of the Secretary-General of the International Seabed Authority under Article 166, paragraph 4 of the United Nations Convention on the Law of the Sea, ISBA/9/A/3, 4 June 2003. As will be seen below, Canada and the US have issued and continue to issue several exploration licenses for the extended continental shelf.

[2] Official text in *The Law of the Sea: Official Text of the United Nations Convention on the Law of the Sea* (New York, United Nations, 1983), Article 76. The area beyond 200M and up to the seaward limits of the continental margin is referred to in this paper as the "extended continental shelf."

[3] A. Chircop and B. Marchand, "International Royalty and Continental Shelf Limits: Emerging Issues for the Canadian Offshore," paper presented at the Canadian Petroleum Law Foundation, Second East Coast Seminar, 18-20 September 2003, Newfoundland and Labrador, Canada, forthcoming in the *Dalhousie Law Journal* in 2004.

[4] There are various regional definitions of the Arctic, and the states identified in this paper are members of the Arctic Council. See D. VanderZwaag, "Regionalism and Arctic Marine Environmental Protection: Drifting between Blurry Boundaries and Hazy Horizons," in: Vidas and Østreng (eds.), *Order for the Oceans at the Turn of the Century* (The Hague, Kluwer Law International, 1999), pp. 231-248, at 231-235.

[5] The US Senate Foreign Relations Committee has recommended that the US Senate give its advice and consent to accession to UNCLOS and ratification of the Part XI Implementing Agreement. Report of the US Senate Committee on Foreign Relations on the United Nations Law of the Sea Convention (11 March 2004). See ASIL International Law in Brief, 23 April 2004. Similarly, The US Commission on Ocean Policy produced a preliminary report advocating accession. *Lloyd's List*, 27 April 2004.

[6] UNCLOS, supra note 2, Art. 82.

[7] *Ibid.*

[8] For an in-depth discussion of the diplomatic history of Art. 82, see Chircop and Marchand, supra note 2; S. N. Nandan and S. Rosenne (vol. eds.), *United Nations Convention on the Law of the Sea 1982: A Commentary*, Vol. 2 (Dordrecht, Martinus Nijhoff Publishers, 1993), commentary on Part VI, Art. 82, pp. 930-947. See also the various UNCLOS III negotiating texts: Art. 69, Informal Single Negotiating Text, ISNT/Part II, A/CONF.62/WP.8, 7 May 1975, in Platzöder (ed.), *Third United Nations*

Conference on the Law of the Sea: Documents, (New York, Oceana, 1983-1988), Vol. I, p. 31; Art. 70, Revised Single Negotiating Text, Part II, A/CONF.62/WP.8/Rev. 1/Part II, 6 May 1976, in Platzöder, *Ibid.*, Vol. I , p. 218; Art. 82, Informal Composite Negotiating Text, A/CONF.62/WP.10, 15 July 1977, in Platzöder, *Ibid.*, Vol. I, p. 318; Art. 82, Informal Composite Negotiating Text/Revision One, ICNT/Rev. 1, A/CONF.62/WP.10/Rev. 1, 28 April 1979, in Platzöder, *Ibid.*, Vol. I, p. 423; Art. 82, Informal Composite Negotiating Text/Revision Two, ICNT/Rev. 2, A/CONF.62/WP.10/Rev. 2, 11 April 1980, in Platzöder, *Ibid.*, Vol. II, p. 51; Art. 82, Draft Convention on the Law of the Sea (Informal Text), A/CONF.62/WP.10/Rev. 3, 22 September 1980, in Platzöder, *Ibid.*, Vol. II, p. 227; Art. 82, Draft Convention on the Law of the Sea, A/CONF.62/L.78, 28 August 1981, in Platzöder, *Ibid.*, Vol. II, pp. 409-410.

[9] "Ocean Space, Seabed, Common Heritage of Mankind," First Statement to the First Committee of the General Assembly, November 1 1967, in A. Pardo, *The Common Heritage: Selected Papers on Oceans and World Order 1967-1974* (Malta, Malta University Press, 1975), pp. 1-41.

[10] Pardo, "From Seabed Regime to Ocean Space Regime," 23 March 1972, in Pardo, *Common Heritage*, *Ibid.*, p. 219. See also Art. 61 of "Draft Ocean Space Treaty," Working Paper submitted by Malta, A/AC.138/153, *Ibid.*, pp. 387-388 and 421.

[11] For an overview of the debate and alternative proposals, see Nandan and Rosenne, supra note 8, pp. 930-947.

[12] For an overview of political theory on international taxation see M. J. Frankman, "International Taxation: The Trajectory of an Idea from Lorimer to Brandt," 24 *World Development* (May 1996, 2000), online: <http://www.globalpolicy.org/socecon/glotax/general/loribran.htm> (accessed 16 April 2004).

[13] *Black's Law Dictionary* (St. Paul, Minn., West Publishing, 1979), p. 1307. Other charges comprehended by "tax" are toll, tribute, tallage, gabel, impost, duty, custom, excise, subsidy, aid, supply, etc. *Ibid.*

[14] E.g., UNCLOS, supra note 2, Article 162(2)(p) and Annex III, Art. 13.

[15] U. Fastenrath, in R. Bernhardt (ed.), *Encyclopedia of Public International Law*, Vol. 4 (Amsterdam, Elsevier, 2000), pp. 387-390, at 390.

[16] Servitude corresponds to the concept of easement at common law, except that servitude refers to the burden, whereas easement refers to the benefit enjoyed. *Black's Law Dictionary*, supra note 13, p. 1229.

[17] "It follows from the nature of a servitude as being bound to a specific territory, that only exceptional restrictions on territorial sovereignty are meant, and not those general restrictions on territorial supremacy which, according to certain rules of international law, apply equally to all states." Fastenrath, supra note 15, p. 388.

[18] *Black's Law Dictionary*, supra note 13, p. 1195. The US Minerals Management Service (Gulf of Mexico OCS Region) defines royalty as follows: "A share of the minerals produced from a lease; a percentage of production either in money or in kind that a Federal Lease is required to pay." Outer Continental Shelf: Oil and Gas Leasing Procedures Guidelines, OCS Report MMS 2001-076, New Orleans, October 2001, p. 85, online: <http://www.gomr.mms.gov/homepg/whatsnew/techann/2001-076.pdf> (accessed 28 April 2004).

[19] On the role of the ISA in this regard, see M. W. Lodge, "The International Seabed Authority: Its Future Directions," paper presented at the Conference on Legal and Scientific Aspects of Continental Shelf Limits, Reykjavik, Iceland, 25-27 June 2003, pp. 403-409.

[20] See Chircop and Marchand, supra note 3.

[21] There have been several licenses and accompanying Lease Stipulations for exploration of the US extended continental shelf, the most recent of which is Lease Stipulations, Oil and Gas Lease Sale 192, Western Gulf of Mexico, Proposed Notice of Sale (US Department of the Interior, Minerals Management Service, Gulf of Mexico OCS Region), Stipulation No. 4: Law of the Sea Convention Royalty Payment (hereafter Lease Stipulations 192). See also MMS Issues Proposed Notice of Western Gulf Lease Sale 192,GOMR Release No. 3039, 26 March 2004. Sale 192 is scheduled for 18 August 2004.

[22] *Ibid.*, Lease Stipulations 192, Stipulation No. 4, para. 1. It is pertinent to note that in the US submerged lands outside the territorial sea are unequivocally federal.

[23] *Ibid.*, para. 5.

[24] Statement by P. L. Kelly, Senior Vice President, Rowan Companies, Inc. on behalf of the American Petroleum Institute, International Association of Drilling Contractors and National Ocean Industries Association before the US Senate Committee on Foreign Relations' Hearing on the United Nations Convention on the Law of the Sea, Washington, D.C., 21 October 2003. Online: <http://foreign.senate.gov/testimony/2003/KellyTestimony031021.pdf> (accessed 16 April 2004).

[25] Kelly, *Ibid.*

[26] Lease Stipulations 192, supra note 21, para. 9.

[27] Informal Single Negotiating Text (ISNT), 1975, Art. 69, in Platzöder, supra note 5, p. 31.

[28] UNCLOS, supra note 2, Arts. 62, 69-72.

[29] It should be remembered that the resources of the continental shelf that are subject to sovereign rights include sedentary species of living resources (e.g., lobster, crab, scallops). UNCLOS, *Ibid.*, Art. 77(4).

[30] UNCLOS, *Ibid.*, Art. 140(1).

[31] In UNCLOS there does not appear to be a duty on the broad margin state to consult the ISA on the development of transboundary resources, although there may be one at customary law. There is, however, a duty on the ISA to act with due regard to the rights and legitimate interests of the broad margin state. The duty requires consultation, including prior notification. See UNCLOS, *Ibid.*, Art. 142.

[32] UNCLOS, *Ibid.*, Art. 153.

[33] Chircop and Marchand, supra note 3.

[34] Provision VIII, Informal Working Paper No. 4/Rev. 2, 27 August 1974, Formula A: "1. All States deriving revenues from the exploitation of the non-living resources of the … zone shall make contributions to the international authority at the rate of … per cent of the net revenues. 2. The international authority shall distribute these contributions on the same basis as the revenues derived from the exploitation of the international sea-bed area." Platzöder, supra note 8, Vol. III, p. 362

[35] M. H. Nordquist and C. Park, *Reports of the United States Delegation to the Third United Nations Conference on the Law of the Sea* (Honolulu, Law of the Sea Institute, University of Hawaii, 1983), pp. 98-100.

[36] See Nandan and Rosenne, supra note 8, pp. 956-947.

[37] The US had proposed that the payments and contributions could be directed at a UN body or regional economic organization. Nandan and Rosenne, *Ibid.*, p. 941.

[38] Final Act of UNCLOS III, Annex I, Resolution I: Establishment of the Preparatory Commission for the International Sea-Bed Authority and for the International Tribunal for the Law of the Sea, UNCLOS, supra note 2, pp. 175-176.

[39] For example with pioneer investor states and entities specified in the Final Act of UNCLOS III, Resolution II: Governing Preparatory Investment in Pioneer Activities relating to Polymetallic Nodules, UNCLOS, supra note 1, pp. 177-182. The ISA and its

subsidiary bodies are bound by the decisions of the PrepCom under this Resolution, *Ibid.*, Art. 14, p. 182.

[40] "With respect to the proposal to carry out a study of the implications of article 82, paragraph 4, of the Convention, some delegations expressed the view that the responsibilities of the Authority under article 82 were strictly limited to the functions set out in article 82, paragraph 4, and that any study by the secretariat should be focused accordingly." As reported in "Statement of the President on the Work of the Assembly at the Ninth Session," Assembly, Ninth Session, Kingston, Jamaica, 28 July-8 August 2002, ISBA/9/A/9, 7 August 2003. See also "Seabed Assembly Discusses Secretary-General's Annual Report," ISA Press Release, SB/9/12, 5 August 2003. The states concerned were Australia, the Russian Federation, U.K. and US and they insisted that the ISA's studies or other activities on the extended continental shelf must be clearly limited.

[41] Nandan and Rosenne, supra note 8. See especially the discussion at the Sixth Session at UNCLOS III, p. 940 et seq.

[42] Presumably pursuant to Regulation 5.5 of the Financial Regulations of the International Seabed Authority, which authorises the ISA Secretary-General to establish trust funds and special accounts. It is not clear how Article 82 revenues will be classed. Regulation 7.1 simply states that Article 82 payments will not be classed as miscellaneous income for credit to the general administrative fund. Online: http://www.isa.org.jm/en/publications/Financial%20Regulations.pdf (accessed 28 April 2004).

[43] UNCLOS, supra note 2, Art. 160(1).

[44] *Ibid.*, Art. 162(1).

[45] *Ibid.*, Art. 162(2)(o)(i).

[46] *Ibid.*, Art. 160(2)(f)(i).

[47] In certain circumstances, it may be possible for the coastal State to argue that there is no duty under UNCLOS, even with reference to the fulfillment of a conventional obligation, to disclose information which might be contrary to the essential interests of its security. *Ibid.*, Art. 302. There is an argument that the ISA may have to monitor activities on the extended shelf in the case of a joint development zone that includes, with its consent, the adjacent International Seabed Area.

[48] This scenario is not as farfetched as may appear. In the case of contributions to the International Oil Pollution Compensation Fund, several states have not provided to the Fund the basic information required to determine contributing oil for the purpose of assessment of contributions to the Fund. *International Oil Pollution Compensation Funds: Annual Report* 2002, p. 27.

[49] UNCLOS, supra note 2, Art. 300.

[50] *Ibid.*

[51] See Informal Single Negotiating Text II (ISNT II), Art. 69(5) proposed by the group of Land-Locked and Geographically Disadvantaged States: "If a State concerned fails to comply with the provisions of this Article the International Authority may take appropriate measures in accordance with the powers and functions conferred upon it by this Convention." Platzöder, supra note 8, Vol. IV, p. 327.

[52] UNCLOS, supra note 2, Art. 291(2).

[53] *Ibid.*, Art. 187 and Annex VI, Art. 37.

[54] *Ibid.*, Art. 187.

[55] *Ibid.*, Art. 189. "Its jurisdiction in this regard shall be confined to deciding claims that the application of any rules, regulations and procedures of the ISA in individual cases would be in conflict with the contractual obligations of the parties to the dispute or their obligations under this Convention, claims concerning excess of jurisdiction or misuse of power, and to claims for damages to be paid or other remedy to be given to the party concerned for the failure of the other party to comply with its contractual obligations or its obligations under this Convention." *Ibid.*

[56] *Ibid.*, Arts.160(2)(f)(i) and 162(2)(o)(i).

[57] *Ibid.*, Art. 285.

[58] M. H. Nordquist, S. Rosenne and L.B. Sohn (eds.), *United Nations Convention on the Law of the Sea 1982: A Commentary*, Vol. V (Dordrecht, Nijhoff, 1989), p. 36.

[59] UNCLOS, supra note 2, Art. 280; Nordquist et al., *Ibid.*, pp. 35-36.

[60] UNCLOS, supra note 2, Annex VI, Arts. 20(1), 21 and 22. See also Art. 288 in the main text of the Convention.

[61] *Ibid.*, Art. 284 read against Art. 285. Art. 285 extends the application of Part XV, Section 1, to entities other than state parties. The reference to "party" in Annex V on Conciliation, Art.1, is not restricted to state parties. Although arbitration under Annex VII does not seem to apply to non-state entities (because Art. 285 does not apply to Part XV, Section 2), in reality Section 1 allows disputants (states and entities) to choose their own dispute settlement procedures. *Ibid.*, Art. 280.

PANEL VI:
NEW CHALLENGES IN THE ARCTIC REGION

Introduction to New Challenges in the Arctic Region

*Sverrir Haukur Gunnlaugsson**

Iceland has served as Chair of the Arctic Council for the past two years, but in fall 2004 Russia will take over the Chairmanship. The Law of the Sea Institute of Iceland therefore found it appropriate to propose and sponsor a special panel at this Conference dealing with some of the new challenges in the Arctic Region. It should be pointed out that in addition to this panel, Panel VII will touch upon an issue that has been the subject of a major study by the Arctic Council, namely climate change.

The Arctic Council, which was established in 1996, is a regional intergovernmental forum for sustainable development, mandated to address all three of its main pillars: environmental, social and economic. The Member States of the Arctic Council are the Nordic countries, Denmark, Finland, Iceland, Norway and Sweden, and Canada, the Russian Federation and the United States.

The scientific work of the Arctic Council is carried out in five expert working groups focusing on such issues as monitoring, assessing and preventing pollution in the Arctic, climate change, conservation and sustainable use of biodiversity, emergency preparedness and prevention, in addition to the living conditions of the Arctic residents.

Since the establishment of the Arctic Council, Arctic governments and indigenous peoples have joined together in making the monitoring and assessment of the Arctic environment a key element of the Council's agenda. Groundbreaking reports on pollution risks and their impact on the Arctic ecosystem and on the conservation of biodiversity have attracted global attention to the state of the Arctic environment. This important work is performed under the Arctic Monitoring and Assessment Programme (AMAP).

The Sustainable Development Working Group (SDWG) has started developing the economic, social and cultural aspects of sustainable development. The Arctic States have declared their commitment to improving human conditions in the Arctic and to building capacity to help the inhabitants adapt to new realities. Recently, the Council began to

* Ambassador of Iceland to the United Kingdom, Chairman of Board, Law of the Sea Institute of Iceland.

produce an Arctic Human Development Report (ADHR), a comprehensive assessment of human conditions in the circumpolar region, which should be completed in fall 2004.

Protection of the Arctic Marine Environment (PAME) addresses policy and non-emergency pollution prevention and control measures related to the protection of the Arctic marine environment from land and sea-based activities, including marine shipping, offshore oil and gas development, land-based activities and ocean disposal. Recognizing that existing and emerging activities in the Arctic warrant a more coordinated and integrated strategic approach to address the challenges of Arctic coastal marine environments, Arctic Council Ministers requested the PAME Working Group to lead the development of an Arctic Marine Strategic Plan (AMSP). The purpose of the AMSP is to guide Arctic Council activities related to the protection of Arctic seas.

Conservation of Arctic Flora and Fauna (CAFF) aims to promote the conservation of biodiversity and the sustainable use of living resources. The working group has published a substantive overview report on biodiversity and conservation in the Arctic, including marine areas.

The Emergency, Prevention, Preparedness and Response (EPPR) working group exchanges information on best practices for preventing spills, preparing to respond to spills should they occur, and practical response measures in the event of a spill. EPPR is not a response agency. The work is focused mainly on oil and gas transportation and extraction, and on radiological and other hazards. Expanded use of natural resources (oil, gas, and mining) and growth in tourism will lead to new and more frequently used navigation routes. This calls for new efforts to enhance the security of marine transport, and prevent emergencies or respond to them effectively, which includes smooth cross-border assistance among neighboring states. The EPPR working group has developed a number of tools, including an Environmental Risk Analysis of Arctic Activities, a Circumpolar Map of Resources at Risk from Oil Spills in the Arctic, a Field Guide for Oil Spill Response in Arctic Waters, and Source Control Management approaches for selected facilities. A project attracting great attention is the ongoing assessment of the impact of climate change in the Arctic. This will be the first comprehensive, regionally based study of climate change to be published since the United Nations Convention on Climate Change in 1992. With temperatures in the Arctic rising at twice the global average, climate change will have an impact on every aspect of life in the Arctic in coming years and decades. Circulation of the atmosphere and the ocean, the biosphere, infrastructure, livelihoods and human health will all be affected to some degree. Understandably, the scientific results of the assessment, due to be published in fall 2004, are eagerly awaited.

Consequences of Rapid Arctic Climate Changes

Alexander Klepikov, Alexander Danilov and Victor Dmitriev[*]

It is known that the Arctic is an important part of the global climatic system, which is connected with its other components by the heat, moisture and water transfers that circulate through the atmosphere and the ocean. During the last two decades of the 20th century, a noticeable global climate warming occurred. In the opinion of many experts the warming will continue throughout the 21st century. At the same time, intense anthropogenic impacts on the environment of the northern regions have been observed, which also influences both the nature of the warming and, indirectly, the climatic system.

The review of the Arctic pollution problems within the framework of the Arctic Monitoring and Assessment Program indicates [AMAP, 2002] that the natural variability of the Arctic is aggravated by the impact of global climatic changes related to the increasing concentration of greenhouse gases, their source, in particular, being combustion of fossils fuels. The International Group of Experts on Climate Change considers that "as a result of anthropogenic activity . . . some climate changes already occur." In particular, the ice cover of the Arctic Ocean as one of the most typical indicators of climatic changes "has undergone some specific changes from the 1990s" – the sea ice has decreased [AMAP, 2002].

The long-term climate changes and the natural climatic cycles determine in many respects the transport of pollutants to the Arctic and within it. This plays one of the major roles for the sustainable development of the region.

Some scenarios of the development of climate change consequences predict an unprecedented development of catastrophic social events even up to the beginning of inter-state armed conflicts, mass migration of the population, and the like. The "extreme" scenario published recently by the U.S. Department of Defense is such an example [Schwartz and Randall, 2003]. However, without understanding the mechanism of global climate formation it is impossible to predict reliably the current changes of weather conditions in any region of the globe.

---

[*]Arctic and Antarctic Research Institute, St. Petersburg, Russia.

The Arctic environmental features present a natural factor, often enhancing the adverse aspects of ecological and social changes. Its own changes can have serious implications and hence require careful attention.

The exclusive importance of the Arctic for Russia in the 21st century is noteworthy. The geopolitical changes in the late 20th century moved the center of the country to the area of intersection of the Polar Circle and the Yenisey River. New Russia is a sub-Arctic nation. The enormous natural resources of the Arctic region enhance its significance even more. An extensive sea boundary and a vast offshore zone with substantial supplies of hydrocarbon raw materials determine the acuteness and specifics of the problems of polar water areas of Russia.

The effective national and international legal systems regulate a wide range of problems and activities. This includes delimitation of sea boundaries and the problem of the outer boundary of the continental shelf. Economic activity covers branches, transport, power engineering, and related matters. Environmental protection is important for the unique and vulnerable nature of the Arctic. Human rights must be considered, including the rights of the indigenous peoples of the North. The existing legislative base regulates the development of offshore resources and bio-resources, environmental protection and reconstruction, life activity of the indigenous peoples and natural complexes, transportation systems, and related issues. Legal regulation inevitably takes into account the peculiarities of the Arctic natural-climatic conditions.

The environmental problems or challenges can be divided into two groups:

- natural, which include natural and indirectly anthropogenic causes – the best example being climate; and
- anthropogenic, which are directly determined by human activity, for example, environmental pollution.

The anthropogenic challenges are more controlled and that is why there are possibilities to mitigate significantly or eliminate their adverse consequences. We shall consider below the problem of natural challenges from the standpoint of the Arctic environment.

The Third Assessment Report of the Intergovernmental Group of Experts on Climate Change published in 2001, states that the most important characteristic of the Earth's climate in the 20th century is an unprecedented increase of global air temperature for the last millennium in the lower layers of the atmosphere comprising about 0.6ºC. According to model calculations, the global climate warming should be stronger in the Earth's polar regions stronger, being especially in the Arctic.

The studies carried out at the University of Alaska showed the winter air temperatures in North America and North Eurasia increased over 30 years (1966-1995) by 2ºC. Simultaneously the summer temperatures in the Arctic have almost not changed. This tendency in the projections of climate experts will continue during the next decades.

An analysis of observations confirms the development of warming in the Arctic beginning from the late 1970s and increasing to the 1990s [Alekseev, 2003]. The mean annual surface air temperature increased on average during this period approximately by 0.5°C over the entire Arctic. This is, however, less than the temperature increase during the period of Arctic warming in the 1920-1940s period and is close to the increase of average temperature of the entire Northern Hemisphere. The area of Arctic sea ice from the late 1970s decreased more rapidly compared to the 1920-1940s, reaching in the first half of the 1990s its least value over a century. Quite significant changes in the 1990s were manifested in the Atlantic water temperature in the Arctic Basin. These increased everywhere up to 0.5-1.5°C reaching values not recorded during the entire observation period from the end of the 19th century.

Warming in the Northern Hemisphere in the second half of the 20th century was most pronounced in the late 1980s – second half of the 1990s, when the highest air temperatures over the period of instrumental observations were recorded. For high latitudes, this temperature increase was not as significant as in the 1930–1940s. However, from the mid-1960s, the temperature of the Northern polar area (60–90° N) has a clear tendency towards an increase. The character of changes in the air temperature and in other meteorological characteristics in different regions of the Northern polar area was not the same [Alekseev, 2003].

According to the results of AARI studies, the anomaly of mean annual air temperature of the Northern Hemisphere in 2002 was 0.66°C. The year 2002 for the Northern Hemisphere is the second warmest year after the warmest year of 1998 [Alekseev, 2003].

The studies indicate the ambiguous and complicated nature of current climate changes in the Arctic. The results cannot be reduced solely to anthropogenic impacts. Further studies are needed using extensive data on paleo-climatic reconstructions of climates in the past and historical climatic data sets up to modern measurements from satellites and other observation platforms operating in the Arctic.

The warming at high latitudes is capable of provoking emissions of natural greenhouse gases accumulated in permafrost which is comparable with the industrial emissions of the developed countries.

In the opinion of the experts, the global air temperature throughout the 21st century will continue to increase with the anticipated further warming in the Arctic region, especially in winter, and the increased annual precipitation totals. According to calculations, climate warming occurs due to the increasing concentration of greenhouse gases in the atmosphere. However, the question remains open as to the extent this is governed by real physical processes and feedbacks in the climatic system.

Carbon dioxide is the main greenhouse gas of the atmosphere. Its concentration increase is responsible for the current and future global climate warming. The results of numerical experiments indicate that if the existing carbon dioxide increase rates are preserved, the mean global air temperature in the surface layer will increase by 2-4 degrees. Thus there will be significant changes in the general circulation of the atmosphere that determine the heat and moisture redistribution over the entire globe. According to some scenarios, the temperature of the coldest month in North Siberia by 2025 will increase by 3-5°C. The same experiments point to a substantial decrease of sea ice areas in the Arctic.

The warming of the air atmosphere at high latitudes predicted for the 21st century will have serious economic consequences. A question arises of whether there could be such natural environmental changes. A need may exist for changes in international and Russian law.

It is worth noting that the melting of glaciers leads to sea level rises. This influences the coastal infrastructure and activities (ports, structures near the waterline, and the like), changes the coastline contours and, as a result, changes the configuration of the territorial sea boundary.

Sea ice decreases and milder ice conditions will also make the Arctic shelf resources more accessible. At the present time, Russia possesses 4.1 million km^2 of the Arctic shelf, which may contain 15-20 billion tons of conventional fuel. Change of the ice regime, displacement of the sea ice marginal zones will influence the marine ecosystems, including commercial objects. New problems of quotas and protection of resources will necessitate changes in regulations.

At present, Russia has full sovereignty of the Northern Sea Route. The UN Convention on the Law of the Sea envisages the existence of special regions within the 200-mile exclusive zones where, due to the recognized technical reasons (oceanographic conditions, ecological aspects), it is permissible to undertake special measures to prevent pollution from ships. In addition, article 234 emphasizes the "ice-covered regions" (where ice is observed for more than six months), where it is possible to expand the prerogatives of the coastal States in the Arctic regions. The peculiarities of the Russian rules of navigation along the Northern Sea Route are based on these provisions. Similar arguments are also used for the Canadian Sea Route – the Northwest Passage. A decrease of the sea ice in the Arctic Ocean and opening of the Northern Sea Route may cause serious juridical problems.

The aforementioned circumstances determine the need for systematic studies of the Arctic. Monitoring of the anthropogenic and non-anthropogenic environmental impacts and assessment of the impact of human activity on the nature of polar regions are urgent objectives of Arctic investigation in the 21st century.

References

1. Alekseev G.V., 2003: Studies of Arctic climate changes in the 20th century. AARI Proceedings., Vol. 446, St. Petersburg, Gidrometeoizdat, p. 6-21.

2. AMAP, 2002: Arctic pollution 2002: Persistent organic pollutants, Heavy metals, Radioactivity, Human health, Changing pathways. Arctic Monitoring and Assessment Program (AMAP), Oslo, Norway, 112 pp.

3. Schwartz P. and Randall D., 2003: An Abrupt Climate Change Scenario and Its Implications for United States National Security. <http://www.fas.org/irp/agency/dod/schwartz.pdf>

PANEL VII: LEGAL AND SCIENTIFIC ASPECTS OF THE NORTH SEA PASSAGE

Climate Change and the Northern Sea Route: An Icelandic Perspective

Thor Edward Jakobsson, Ph.D. *

I. Nature, Environment and Climate Research

a) The Sun

The sun is the supreme ruler of our solar system, worshipped by our ancestors and admired by mankind for thousands of years until this very day. Inside the sun, energy is created and, without interruption, is thrown out into space in all directions. In a few minutes up to a few hours some of this energy reaches the relatively small planets orbiting the mighty sun. Among those planets is Earth, receiving a tiny fraction of the solar energy stream but still enough to heat the surface, create atmospheric and oceanic movements, and create suitable conditions for life on Earth.

b) Global Climate

Climate on Earth is variable from one place to another. The great variety is due to different exposure to the sun. The Earth happens to rotate around itself in a plane that is different from the one it revolves around the sun. As a result, a specific location on Earth phases the sun more directly at one time of the year than it does during another time. Thanks to the angle between the plane of daily rotation and the Earth's orbit around the sun, we enjoy the sequence of the seasons coming and going one after another.

Latitude, or the distance from the equator, is a ruling parameter defining climate on Earth. Another important parameter in defining climate is the geographical distribution of continents on the globe. This dominates the course of ocean current systems and affects the general atmospheric flow crossing the mountain ridges. It results in variable surface conditions, which help in shaping the nature of local air masses.

---

* Project Manager, Sea Ice Unit, Weather Prediction Department, Icelandic Meteorological Office, member of Expert Team on Sea Ice/Joint Technical Committee of Oceanography and Marine Meteorology/World Meteorological Organization.

Climatologists have put much effort into classifying Earth's climate, defining continental and oceanic climates, and combinations of these, as well as the variability in time of the climatological parameters. Conventionally, a specific location's climate is defined by its average conditions over 30 years, sometimes taking into account variability.

c) Climate Variations

For a considerable amount of time it has been known that climate has changed in the past. There have been periods with different glaciation, indicating great variations in temperature and rainfall. Research in the last 50 years has added to this knowledge of variations on shorter time scales of centuries and even decades. Climate, particularly at high latitudes, seems to be changing all the time, just as the weather is changing all the time on a shorter time scale.

d) Global Warming

Investigations of long records of temperature, rainfall and other meteorological observations since the 19th century have demonstrated that climate has been slowly changing from one decade to another. In northern latitude, conditions were generally rather cold until about 1920, when a slightly warmer period started. It lasted until approximately 1965, when a somewhat cooler period came about. Toward the end of the 1970s a gradual change occurred again and, in general, during the last quarter of the 20th century and the first years of the 21st century, climate was characterized by relatively mild conditions.

e) Causes

Much research has been done to determine possible causes of short-term climate variations. Various explanations have been proposed, including the following: The variations could be caused by volcanic eruptions, variable solar energy, atmophere/ocean interaction, internal atmospheric fluctuations or effects due to industry, i.e., pollution. Combinations of these effects have also been suggested, and their comparative strength discussed and studied.

f) Climate Modelling

The advent of modern-day computers has made modelling of natural phenomena possible. Numerical weather predictions were made soon after World War II and climate modelling research started a couple of decades later. Complicated models have been designed where the interaction between the oceans and the atmosphere and other phenomena of the climate system can be studied. Experiments have been done with various amounts of greenhouse gases. The warming effect of increasing amounts of these gases in the atmosphere has been estimated by many research groups. In general, a warming trend is predicted to occur in the 21st century, the extent of which is being debated.

Global climate models generally agree that warming due to increased greenhouse gas concentrations is predicted to be most intense in high latitudes. For instance, the GFDL (Geophysical Fluid Dynamics Laboratory) global climate model predicts a 100-year temperature change of about 3 degrees Kelvin globally and 5 degrees Kelvin near the poles. This is largely because melting sea ice changes the albedo of high-latitude oceans, and to a lesser extent because an inversion prevails at high latitudes, especially in winter, whereas at low latitudes the heating is convectively mixed throughout the troposphere.[1]

II. The Arctic Ocean and Sea Ice

It is common knowledge that the environmental conditions around the two poles – the cold sinks on Earth – are very different. Antarctica is one of the continents with huge glaciers reaching into the Southern Ocean surrounding Antarctica, where extended sea ice also forms in winter. On the other hand, around the northern pole presence of an ocean, the Arctic Ocean creates the environment, characterized by uniqe weather, ocean currents and sea ice climate. In the next couple of sections the nature of the Arctic Ocean will be described. The description is based on references (2) and (3).

a) The Geography and Climate

Nearly landlocked, the Arctic Ocean is bordered by Greenland, Canada, Alaska, Russia, and Norway. The Bering Strait connects it with the Pacific Ocean, and the Greenland Sea is the chief link with the

Atlantic Ocean. The floor of the Arctic Ocean is divided by three submarine ridges—Alpha Ridge, Lomonosov Ridge, and the Arctic Mid-Oceanic Ridge; other submarine ridges, such as the Faeroe-Icelandic Ridge, act to separate the Arctic Ocean from the Atlantic.

The Arctic Ocean is located entirely within the Arctic Circle and occupies the region around the North Pole. It is approximately 14 million square kilometers (km), slightly more than 1.5 times the size of the United States and smallest of the world's four oceans, after the Pacific Ocean, Atlantic Ocean and Indian Ocean. It includes Baffin Bay, Barents Sea, Beaufort Sea, Chukchi Sea, East Siberian Sea, Greenland Sea, Kara Sea and Laptev Sea.

The Arctic Ocean has the widest continental shelf of all of the oceans; it extends approximately 1,200 km seaward from Siberia. Numerous islands rise from the shelf, including the Arctic Archipelago, Novaya Zemlya, the New Siberian Islands and Wrangel Island. The continental shelf encloses a deep oval basin (with an average depth of almost 3.7 km) that stretches between Svalbard and Alaska; east of Greenland the ring of the continental shelf is broken by the Greenland Sea. The greatest depth (approximately 5.5 km) in the Arctic Ocean is found just north of the Chukchi Sea.

The climate of the Arctic is characterized by continuous cold and relatively narrow annual temperature ranges. Winters are characterized by continuous darkness, cold and stable weather conditions, and clear skies. In summer there is continuous daylight, damp and foggy weather, and weak cyclones with rain or snow.

The central ocean surface is covered by a perennial drifting polar icepack that averages about three meters in thickness, although pressure ridges may be three times that size. It has a clockwise drift pattern in the Beaufort Gyral Stream, but nearly straight line movement from the New Siberian Islands (Russia) to Denmark Strait (between Greenland and Iceland). The ice pack is surrounded by ocean seas during the summer, but more than doubles in size during the winter and extends to the encircling land masses.

b) Oceanography and Environment

Since the Arctic's connection with the Pacific Ocean is narrow and very shallow, its principal exchange of water is with the Atlantic Ocean through the Greenland Sea. Even there, though surface waters communicate freely and a strong subsurface current brings warm water from the Atlantic into the Arctic basin, exchange of deeper waters is barred by submarine ridges. Thus a near stagnant pool of very cold water is found at the bottom of the Arctic basin.

Because several major rivers in Siberia (Lena, Yenisei, Ob) and Canada (Mackenzie) bring in much water, and because evaporation is only slight, the outflow through the Greenland Sea is important. It creates the cold East Greenland Current, which flows south along the coast of East Greenland. A weaker current goes through Smith Sound and Baffin Bay and is known as the Labrador Current. Another weak current flows out of Bering Strait. The water that does not flow out by the Greenland Sea seems to be deflected by North Greenland and forms the current that gives rise to a circular current in the Arctic basin itself. This circular current causes the relatively light ice of the Siberian seas, which contrasts with the heavy-pressure ice phenomenon off Greenland and Ellesmere Island (in the Arctic Archipelago). The drift of ice southward and westward has been noted and utilized by explorers.

The Arctic Ocean is covered with ice four to six meters thick throughout the year in most of its central and western portions. Some of the ice pack remains in the Arctic basin, and some, carried out by the East Greenland Current, melts before going far enough south to reach the regular Atlantic shipping lanes. The icebergs that harass ships are generally brought from the fjords of West Greenland by the Labrador Current. It was long thought that no non-oceanic life could exist in the Arctic; however, despite drifting ice, ice packs, vast ice floes, and winter temperatures of 51 degrees below zero Celsius, there are hares, polar bears, seals, gulls, and guillemots as far north as 88°.

The cold Arctic currents give the shores of the northeastern parts of North America and Asia a much colder climate than the northwest shores of Europe and North America, which are warmed by the North Atlantic Drift and the Japan Current. The Arctic currents are also less saline and lighter than these warmer currents, and therefore the Arctic water is at the surface and the Atlantic current beneath, where they are exchanged in the Greenland Sea.

c) Climate Change and Decreasing Sea Ice

Long-term fluctuations in sea ice extent indicate general climate variations. Studies of sea ice observations and measurements in the Arctic during the last half-century have supported the idea that climate undergoes short-term changes. A comparison of data obtained around 1960 with more recent observations has demonstrated both decreasing concentration of extent of sea ice and decreasing thickness of the Arctic ice cover.

In a study of Johannessen, Shalina and Miles,[4] satellite remote sensing data were used. They write in 1999 that there had been about a three percent per decade reduction in the areal extent of the Arctic sea ice cover since 1978, although it is unknown whether the nature of the perennial ice pack has changed. Satellite data were used to quantify changes in the ice cover's composition, revealing a substantial reduction of about 14 percent in the area of multiyear ice and the spatially averaged thickness of the perennial ice pack, which suggests that the satellite-derived areal decreases represent substantial, rather than only peripheral, changes. If this apparent transformation continues, it may lead to a markedly different ice regime in the Arctic, altering heat and mass exchanges as well as ocean stratification.[5] In their conclusion Johannessen et al. state that the balance of evidence indicates an ice cover in transition, which, if continued, could lead to a markedly different ice-ocean-atmosphere regime in the Arctic.

In a comprehensive study of sea ice data in the Nordic Seas, Torgny Vinje has investigated anomalies and trends and compared atmospheric circulation during the period from 1864 to 1998.[6] The area comprises the Greenland, Iceland, Norwegian, Barents and Western Kara Seas. The extent of ice in the Nordic Seas measured in April 2004 had decreased by approximately 33 percent over the past 135 years. Retrospective comparison indicates that the recent decrease in ice extent is within the range of variability observed since the 18th century. Temporal, monotonically reduced extreme events occur with intervals of 12 to 14 years, suggesting that series longer than about 30 years should be considered to obtain statistical significance regarding temporal changes. Otherwise, decadal temperature variation is also found in the northbound warmer ocean currents.

III. SHIPPING IN NORTHERN OCEAN AREAS, ARCTIC SEA ROUTES AND SCIENTIFIC PROJECTS

a) A Few Words on Exploration

The Central Arctic basin was almost wholly unexplored until the Amundsen-Ellsworth flight over it in 1926.[7] Arctic research was stimulated when it was recognized that the shortest air routes between the great cities of the Northern Hemisphere cross the Arctic Ocean. Improved technology has also facilitated research, with the development of aerial and satellite photography and photogrammetry for precise mapping, the sonic echo sounder for measuring ocean depths, and radio to maintain contact with the rest of the world.

Detailed knowledge of drifts and ice floes, water depths, and the ocean floor has vastly increased. Soviet polar scientists in 1948-49 investigated the Lomonosov Ridge, an undersea mountain range that influences the pattern of ice drift and the circulation and exchange of water in the Arctic Ocean. American scientists in 1959 discovered the existence of a submarine plateau rising almost 2.5 km from the ocean floor. In 1995 the U.S. Navy agreed to lend its force of nuclear attack submarines for a series of civilian expeditions to the Arctic.

The fact that the Arctic Ocean is warming is of great potential importance and is now being studied. Recorded temperatures, glacial regressions, and the appearance of observed species of fish in larger numbers, at higher latitudes, at earlier seasons, and for long periods prove that over the decades a "climatic improvement" has taken place. Similar changes have been reported in sub-Arctic latitudes. Whether the warming is a phase in a cycle or a permanent development cannot yet be said.[8]

b) The International Northern Sea Route Programme (INSROP)

A few research groups have been studying the Northern Sea Route quite thoroughly from different angles, considering natural conditions and ice navigation, environmental factors, trade and commercial shipping aspects and political, legal and strategic factors.

The most thorough study of Arctic shipping is the comprehensive, six-year International Northern Sea Route Programme (INSROP), undertaken by Norwegian, Russian and Japanese institutes and organisations from 1993 to 1999. Results and main findings were described in books, symposia proceedings and at least 167 technical working papers (see Kitagawa,[9] Östreng[10] and Ragner[11]).

In the User Conference Proceedings, Captain Lawson W. Brigham deals in his chapter[12] with the natural environment, ice navigation and Arctic ship technology. He writes that a synthesis of INSROP information points to a clear distinction between the eastern and western operating regions of the Northern Sea Route. For key reasons – such as a colder regional climate, more severe ice conditions and a confining geography – the Arctic coastal region from Vil'kitskiy Strait east to Bering Strait is one of the most challenging ship operating environments on the planet. The length of the navigation season in this region has traditionally been July through October, although sea ice reductions during the 1990s (primarily in the East Siberian Sea) have allowed a modest extension of ice navigation through November and, on occasion, into December. A 150-day navigation season is currently feasible. Contrasting with this extreme situation is the effective, year-round Arctic marine operation maintained in Dudinka (across the Kara Sea to the Yenisey River) since the late 1970s. Continued warming in northern Eurasia and throughout the Arctic may, in compelling ways, change this divided picture of the Northern Sea Route.

Brigham concludes by stating the following:

> [T]he Russian maritime Arctic presents unique challenges to marine operators and to modern technology. To summarize, this key part of the INSROP effort integrates a triad of practical factors linked to Arctic marine transportation – knowledge of the polar operating environment, the ice navigation practices used in the region, and the type of commercial ships envisioned for any future ice routes. With this fundamental information readily available, appropriate projections, planning and risk assessment can be accomplished to determine the plausible limits to commercial ship operations along the Northern Sea Route.

In *Proceedings of the Northern Sea Route User Conference*,[13] O.M. Johannessen et al. discuss the effect of recent decreasing sea ice occuring in the Arctic:

> [T]here is growing evidence from observations that the Arctic sea ice cover is undergoing a significant change where both ice area

and ice thickness are decreasing. Also climatic simulation models predict that the sea ice will be reduced significantly in the next 50 years. The reduction has started and is documented by 20 years of satellite data observations. A reduced ice cover will facilitate sea transportation in the Northern Sea Route significantly. Several scenarios can be foreseen: less ice-covered areas will make it possible to use lower ice-class vessels and navigation can be carried out with less icebreaker support; longer melt season will extend the navigation period along the Siberian coast; finally, thinner ice-cover and more open leads in the central Arctic allows powerful icebreakers to sail efficiently across the North Pole, which is the shortest distance between the Barents Sea and the Bering Strait.

In their conclusion Johannessen et al. discuss the implications of decreased sea ice for the Northern Sea Route:

> There are several important consequences of reduced ice for sea transportation in the Northern Sea Route. Firstly, the navigation period can be longer because larger parts of the Siberian coast will potentially be ice-free or covered by thinner ice in summer and autumn. Secondly, the requirements for icebreaker support in the navigation season can be reduced. Thirdly, the risk for damage to vessels and time loss due to difficult ice conditions can be reduced. Finally, the possibility to use the North Pole route, which represents the shortest sailing distance between the Barents Sea and the Bering Strait, can be enhanced as a result of thinner multi-year ice and more open leads in the high Arctic ocean.
>
> The predicted decrease in the Arctic ice cover is valid for averaged, long-term conditions. There will clearly be large regional and inter-annual variations in the ice cover, which implies the ice conditions can in some years become more severe. The current climate models cannot predict regional changes of the ice cover very well. However, the models are improving and it is foreseen that prediction models for the Northern Sea Route will become more reliable for planning and support of sea transportation.[14]

The main conclusion of the International Sea Route Programme (INSROP; 1993-1999) was that in spite of climatic, technological and political restraints, an increase in international commercial shipping *is* feasible – in economic, technological and environmental terms.

c) Arctic Climate Impact Assessment (ACIA)

The Arctic Climate Impact Assessment (ACIA) is an international project of the Arctic Council and the International Arctic Science Committee (IASC) that evaluates and synthesize knowledge on climate variability, climate change, and increased ultraviolet radiation and their consequences.

The Arctic Council is a high-level intergovernmental forum that provides a mechanism to address the common concerns and challenges faced by the Arctic governments and the people of the Arctic. The member States of the Arctic Council are Canada, Denmark, Finland, Iceland, Norway, the Russian Federation, Sweden and the United States.

The ACIA project is in its final stage, and its 1,800-page report by some 300 contributors will be made public in 2004. The report will be handed to ministers during an Arctic Council meeting in Iceland in November 2004. In a Reuter's news announcement on May 21, 2004, Robert Corell, chairman of the ACIA project, said the following concerning matters relating to Arctic shipping: "There is dramatic climate change happening in the Arctic right now, about 2-3 times the pace of the whole globe."

The news item also stated that benefits for human commerce might accrue from the opening up of a now largely icebound short-cut sea route from the Pacific to the Atlantic. Russia might also win easier access to oil and gas as the icecap shrinks and permafrost retreats.

Corell said that the sea route between the Pacific and the Atlantic via the Arctic could open far earlier than expected by most previous studies, cutting shipping times compared to routes via the Suez or Panama canals. "On average our models show that by 2050 the Northern Sea Route will be open about 100 days a year. Now it's open about 20 days," he said.

d) Interest in Iceland: The Reykjavik Arctic Conference in 1987

In October 1987 an international conference on the Arctic Sea Route, sometimes called the Northern Sea Route, was held in Reykjavik, Iceland.[15] It was a culmination of several years of discussion in Iceland among a few interested people. The interest in Iceland can be traced to a 1982 Soviet-Icelandic expedition on the Russian icebreaker *Otto Schmidt* in the Greenland Sea in which the author of this paper participated. On receiving a report on the voyage, Japanese economist Dr. T. Endo wrote in 1984 to the publisher of the report about Japanese scientists' interest in a future Polar Ocean sea route with Iceland playing a main role due to its location. Iceland could become the Hong Kong of the North!

The 1987 conference was organized by the Association of Municipalities in the Capital Area of Iceland. The meeting happened to be held exactly a week after Secretary General of the Soviet Republics Mikhail Gorbachyov gave a groundbreaking speech on international affairs in Murmansk on October 1, 1987. Toward the end of his speech he made the following important statement:

> The shortest sea route from Europe to the Far East and the Pacific Ocean passes through the Arctic. I think that, depending on progress in the normalisation of international relations, we could open the North Sea Route to foreign ships with us providing the services of ice-breakers.

The more or less sea ice-covered Arctic Sea Route connects the two great oceans, the Atlantic and the Pacific. The conference organizers proposed that the Arctic Sea Route might, in the near future, become a major shipping route, and that Iceland, due its geographical location, would then be a natural locality to serve as an entrepôt. Since then the idea has been kept alive in Iceland among the few enthusiasts.

The Northern Sea Route saves up to 40 percent of the distance from Northern Europe to Northeast Asia and the northwest coast of North America, compared with southerly sea routes via Suez or Panama. The route may also prove instrumental in connecting the Russian Arctic with the rest of the world. The most obvious obstacles to commercially viable shipping in the Northern Sea Route are the harsh natural conditions, including ice most of the year.

Comparing sailing distances one could imagine what route would be most traversed *if* there was no sea ice or very limited sea ice floating around in the Arctic Ocean. From Northwest Europe to Northeast Asia, sailing distance in nautical miles via the Suez Canal is approximately 11,500, but via Iceland and Arctic Route it is approximately 7,600. From Northeast North America to Northeast Asia via the Suez Canal it is approximately 13,200, whereas it is only about 8,900 via Iceland and the Arctic Route.

In recent years, due to scientific reports on decreasing sea ice in the Arctic, a revival of interest has developed in Iceland. In a recent preliminary feasibility study by Nigel Chattey Associates, Inc., USA, on the Arctic Sea Route and the role of Iceland as an entrepôt, it was concluded that Iceland certainly has a potential to serve in such a role.[16]

There has also been increased interest in Iceland due to decreasing sea ice in the East Greenland Current and the Denmark Strait and the possibilities of more frequent marine communication between Iceland and East Greenland.

A further development took place in Iceland in 2003, indicating a breakthrough regarding official support of the idea of strengthening Icelandic ties with Arctic shipping. It was the decision of the Icelandic Ministry of Foreign Affairs to establish a technical study group to deal with these fascinating possibilities. At the same time, the working group considers environmental issues and concerns relating to future Arctic shipping. The group was assigned to investigate what role Iceland could play in future traffic across the Arctic Ocean between the northern parts of the North Atlantic and the North Pacific Oceans and, in particular, the further development of the Northern Sea Route.

e) Final Remarks

In this short article great possibilities concerning trans-Arctic shipping have been briefly considered. Vastly increased actvities in the Arctic and the Sub-Arctic should be expected in the near future, thanks to global warming and retreating sea ice. The many sides of this topic have already been discussed in working groups in several countries where technical, economic and environmental issues are evaluated.

In addition to Russian captains and their crews having 100 years of Arctic sailing experience, other factors such as modern technology progress in the fields of ship building, information and communication technology, remote sensing from satellites and computer modelling of weather and sea ice development make it slowly but ever more economically feasible to establish an international route between the northern parts of the North-Atlantic Ocean and the Pacific Ocean across the Arctic Ocean.

This new development has already been noted by respected international organizations, such as the World Meteorological Organisation (WMO). Among the most important departments of the WMO is the Joint Technical Commission of Oceanography and Marine Meteorology (JCOMM). At a meeting of the second session of JCOMM Expert Team on Sea Ice (ETSI) in Hamburg, Germany, April 15-17, 2004, the following remark was made: "Recognizing the likelihood of increased economic activities in the Arctic in the future, the ETSI underlines the importance of sea ice services, standards, observations and data. In this respect, the scope of ETSI activities may expand in the future."[17]

IV. Arctic Sea Routes and the Possible Role of Iceland

As described in this paper, an interest in the possibility of Iceland playing a role in a trans-Arctic sea route was developed more than 20 years ago by Japanese and Icelandic scientists. The belief in this idea stems from Iceland's unique geographical position, being en route along future traffic lines through the Nordic Seas between Greenland and Scandinavia into the Arctic Ocean. These first discussions thus occurred during the USSR era and before the global warming age in research and international politics. The optimistic view of future possibilities of the Arctic sea route was thoroughly based on considerations of scientific progress in remote sensing, information technology, icebreaker design, weather and sea ice forecasting and, last but not least, on ever-increasing knowledge and experience of sailing in the Arctic environment.

The present political climate in a post-Soviet era as well as predictions of warmer climate in nature have opened many eyes in these matters and quickened investigations of the feasibility of making further use of the Arctic sea routes. Fascinating possibilities relating to a friendly and prosperous cooperation between northern countries will ensue.

References:

1. Linacre, E. and B. Geerts, Web site: <http://www-das.uwyo.edu/~geerts/cwx/notes/chao15/polar.html>.

2. Ocean Planet, Smithsonian, Arctic Ocean Geograhpy, Web site: <http://seawifs.gsfc.nasa.gov>.

3. Infoplease, Encyclopedia, Arctic Ocean, Web site: <http://www.infoplease.com>.

4. Johannessen, Ola M., Elena V. Shalina and Martin W. Miles, Satellite Evidence fro an Arctic Sea Ice Cover in Transformation, Science, 3 December 1999, Volume 286, pp. 1937-1939.

5. *Id.*

6. Vinje, T., Anomalies and Trends of Sea-ice Extent and Atmospheric Circulation in the Nordic Seas during the Period 1864-1998, Journal of Climate, 1 February 2001, pp. 255-267.

7. CIA - The World Factbook – Arctic Ocean, Web site: <http://www.cia.gov/cia/publications/factbook/geos/xq.html>.

8. *Id.*

9. Kitagawa, H. (ed.) Northern Sea Route, Future and Perspective, The Proceedings of INSROP Symposium, Tokyo 95, Ship and Ocean Founation, Tokyo 1996, 687 pp.

10. Östreng, W. (ed.) The Natural and Societal Challenges of the Northern Sea Route. A Reference Work. Kluwer Academic Publishers, Dordrecht/Boston/London 1999. ISBN 0-7923-6112-1. 466 pp.

11. Ragner, Claes Lykke (ed.) The 21st Century – Turning Point for the Northern Sea Route?, Proceedings of the Northern Sea Route User Conference, Oslo, 18-20 November 1999. Kluwer Academic Publishers, Dordrecht/Boston/London 2000. ISBN 0-7923-6365-5. 308 pp.

12. Brigham, Lawson W., Natural Conditions, Ice Navigations and Ship Design – Pushing the Limits, in Proceedings of the Northern Sea Route User Conference, Oslo, 18-20 November 1999. Kluwer Academic Publishers, Dordrecht/Boston/London 2000. ISBN 0-7923-6365-5. 308 pp.

13. Johannessen, O.M., M.W. Miles, H. Drange, G. Evensen, K.A. Lisæter and S. Sandven, Arctic Sea Ice Reduction - Implications for the Northern Sea Route, in

Proceedings of the Northern Sea Route User Conference, Oslo, November 18-20, 1999. Kluwer Academic Publishers, Dordrecht/Boston/London 2000.
ISBN 0-7923-6365-5. 308 pp.

14. *Id.*

15. The Arctic Sea Route, A seminar held in Reykjavik, Iceland, October 8, 1987. Association of Municipalities in the Capital Area, Iceland (mimeographed proceedings; 121 pp.).

16. Chattey, Nigel, Arctic Sea Route, Pre-Feasibility Assessment and Proposal for a Demonstration Project including Suggestions for Possible US and Canadian Participation, Nigel Chattey Associates, Inc., New York, June 2003 (mimeographed report; 22 pp.).

17. JCOMM EXPERT TEAM ON SEA ICE (ETSI), Second Session, STEERING GROUP FOR THE GLOBAL DIGITAL SEA ICE DATA BANK (GDSIDB), Tenth Session, Hamburg, Germany, April 15-17, 2004, Final Report, JCOMM Meeting Report No. 28. Joint WMO/IOC Technical Commission for Oceanography and Marine Meteorology, Page 11, Geneve, 2004.

Annex

THE SUN AND VENUS, June 8, 2004

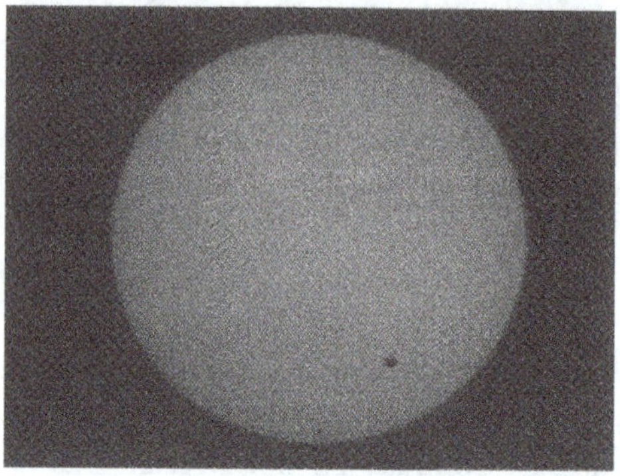

"The Arctic Mediterranean"

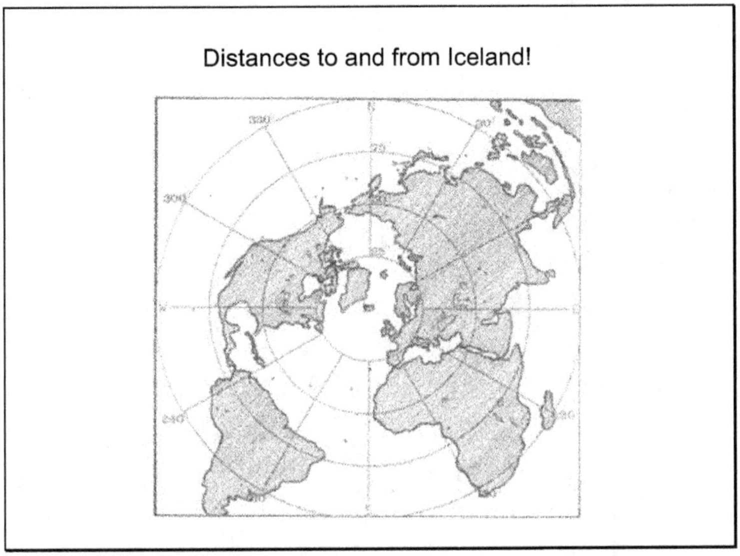
Distances to and from Iceland!

New Illusions of a Northwest Passage

Franklyn Griffiths[*]

Increasingly, those of us who pay attention hear claims that climate change makes intercontinental navigation a certainty in the waters of the Canadian Arctic Archipelago. Specifically, it is said that global warming makes for rapidly thinning ice, indeed, an "ice-free" Arctic Ocean. This, the prediction goes, will soon see the Northwest Passage open wide to commercial shipping between Europe and Asia by non-ice-strengthened vessels during the summer months. I disagree.

Climate change *is* a reality. It will continue to give us global warming unless something happens to make things colder. As well, sea-ice cover *is* thinning in the Arctic Ocean and in the various waterways that make up the Northwest Passage. But summer-months ice conditions now and over the next three decades will continue to vary greatly from season to season, and from region to region within the Canadian Archipelago. In circumstances such as these, the economics of marine transportation between Suez and the Northwest Passage, for example, make it all but certain that major firms will decline to open a new intercontinental shipping route across Arctic North America. And if and when they do get going on a new route between the Pacific and the Atlantic, it is likely to be a transpolar one across the Arctic Ocean, again as summer-months conditions permit. In the meanwhile, an increase in opportunistic use of the Passage by non-ice-strengthened ships on intercontinental voyages is not to be ruled out entirely. More significant is the prospect of year-round North-South navigation to bring hard minerals and, then, liquefied natural gas from Arctic North American locations to southern markets in heavily strengthened ships. This prospect, together with more immediate security concerns connected with illicit Arctic navigation, creates an opportunity for new collaboration between Canada and the United States on Arctic waters issues.

---

[*] Franklyn Griffiths is emeritus professor of political science at the University of Toronto. Comments welcome to franklyn.griffiths@utoronto.ca

The Claim

Numerous sources could be cited for the claim that that the Northwest Passage is becoming a live option for intercontinental navigation. Two stand out. One is the Office of Naval Research (ONR), which, together with other U.S. agencies, sponsored an inquiry into the naval implications of changing Arctic ice conditions as of 2001.[1] The other is a colleague of mine, Professor Rob Huebert of the Centre for Military and Strategic Studies at the University of Calgary, who issued papers on the subject in 2002 and 2003.[2] Whereas ONR and Huebert share a more widespread propensity to exaggerate in forecasting Arctic ice conditions, Huebert goes further in accenting the readiness of shipping firms to run the Northwest Passage. In fact, he likens Canada's business position to that of a straits State: "Singapore has demonstrated that with the proper planning, geographic location on an international strait can bring substantial economic benefits."[3] Similarly, "If it's handled correctly, you sit on an international strait, take a proactive stand and profit nicely."[4] Evidently the Northwest Passage is in for a high volume of intercontinental traffic. A principal problem with this kind of thinking is that a good number of people in Canada happen to take it seriously, especially where ice conditions are concerned.[5] Something similar seems to be going on in the United States.

Entitled "Naval Operations in an Ice-free Arctic," the ONR symposium anticipates a steadily more navigable Arctic as sea-ice extent decreases by three percent per decade. We are told that modeling points to a reduction in sea-ice volume of between 15 and 40 percent by 2050. As well, a summertime disappearance of ice in the Arctic Ocean is posited by 2050.[6] The report also offers a number of more focused estimates. First, there is the suggestion that by mid-century "[t]he Northwest Passage through the Canadian Archipelago and along the coast of Alaska will be ice-free and navigable every summer by non-icebreaking ships."[7] The way things are going, and given that the length of the "summer" is not specified, this projection could well turn out to be correct. Second, and as of 2001, there is the following suggestion: "Within 5-10 years the Northwest Passage will be open to non-ice-strengthened vessels for at least one month each summer."[8] Open end to end for 30 consecutive days every summer without exception beginning at some point between 2006 and 2010? This is surely an exaggeration. The third suggestion: "Within five years, the Northern Sea Route (NSR, a.k.a. the Northeast Passage) will be open to non-ice-strengthened vessels for at least two months each summer."[9] As I will show in some detail, the ONR tends to go overboard in its assessment of future Arctic ice conditions.

Actually, the report itself makes clear that the reference to "ice-free" in its title is misleading. The real focus of attention, we are told, is not an "ice-free Arctic" but "a navigable Arctic with ice-infested waters."[10] ONR is to be commended for its frankness, but still there is quite a difference. Some of the reasons for speaking in terms of an ice-free Arctic are buried in the following bits of bureaucratese from the report: "The…group recognized that energizing DON [Department of the Navy] interest in resourcing operations related to future Arctic operations requires validation and ongoing refinement of existing chronological forecasts."[11] In the same vein, "A more precise forecast that shifts the time window of probable occurrence to the left would present a more compelling argument to DOD/DON leadership for enhanced interest and subsequent planning activities related to the associated contingencies."[12] Translated into English, there are good bureaucratic reasons to exaggerate rates of change and consequences in assessing future Arctic ice conditions. The reasons lie in the phrase "energizing…interest."

Huebert joins the U.S. Navy in speaking of "an ice-free Northwest Passage."[13] He goes on to say that "even for a limited time, [this] would be of tremendous interest to major international shipping companies as well as the countries that avail themselves of their services."[14] If Huebert also aims to "energize interest" in citing the potential for an ice-free Passage, as I believe he does, it is in defence of Canadian sovereignty over the waters of the Archipelago (this notwithstanding his readiness to characterize the Northwest Passage as an international strait). In a view parallel with the ONR, he sees Arctic ice as thinning rapidly now and perhaps even more so in the near future.[15] To these considerations he adds evidence of increased occasional shipping through the Northwest Passage, and signs of new ice-capable vessel construction in Russia and the United States.[16] Further, he points out that use of the Northwest Passage saves some 8,000 kilometers over the Panama Canal between Asia and Europe or the U.S. eastern seaboard.[17] Put all of this together, and in his view it is logical that international shipping interests will want to exploit the opening that is coming by sending vessels through the Archipelago.[18] Since the status of the Passage is disputed in international law, and since transiting vessels may not accept Canadian regulations, the scene could well be set for a challenge to Canadian sovereignty. Indeed, the scene is set as Huebert sees it. The question is not whether it will happen. Instead, he says, "It is impossible to know who will make the first challenge."[19] I regard this way of thinking as mistaken.

Whether it's new Arctic capabilities for the U.S. Navy or new icebreakers and other assets for the defence of Canadian sovereignty, the vision of an ice-free Arctic seems hard to avoid in generating the required interest and securing the necessary appropriations. But the vision is ill-founded. Policy built on an overwrought reading of the situation is bound to fall short of expectation when it does not fail outright. So, let us stow the hype and try to get the situation report right.

Ice Conditions

A lot of people have in-depth knowledge of what has been happening to the ice in the Archipelago. Inuit, from the Inuvialuit on the Beaufort to the people of Nunavut in the area of Lancaster Sound and eastwards, surely have the most knowledge about localities and their cycles. Mariners, especially Canadian icebreaker captains and ice navigators, have acquired a hands-on experience of the Northwest Passage in its entirety. But those with the entire picture in mind year in and year out and in systematic fashion are to be found in the Canadian Ice Service. A unit of Environment Canada, the Ice Service might one day be working with Inuit to fuse "normal" science with traditional ecological knowledge. For now, it is the overarching interpretations of the Ice Service that command our attention.

The key questions have to do with the where, when and what of ice in the Archipelago. "What" refers to first-year and multi-year ice, the former having been generated over the winter and the latter having survived one or more summers. Multi-year ice loses salt and thus becomes very hard. It is the equivalent of floating steel. Brush against a mass of it with a single-hulled, non-ice-strengthened vessel and the result may well be mortal. Or get stowed in and it costs $100,000 and time lost sailing just to get the ship up in dry dock for an inspection.

As to the "where" of ice, the Service reports that it is diminishing everywhere. Throughout Canada's Arctic waters, Hudson Bay included, sea-ice cover has decreased by about 15 percent since 1969.[20] As well, ice cover at the summer minimum has diminished by about 8 percent per decade over roughly the same period in the eastern and western portions of the Archipelago, which is to say in the area of the main waterways of the Northwest Passage.[21] Further, if existing rates of reduction were to persist into the 2060s, the worst summer months would present less ice than the lightest year to date.[22] On the other hand, discontinuities are also to be observed. Within the western Arctic region, for example, rates of reduction vary from 11 percent per decade for the western Arctic waterway, to three percent for the Viscount Melville Area.[23] And while the eastern region shows a reduction rate of five percent per decade, at the eastern entry to the Passage in Lancaster Sound there is no evidence of a decrease in ice cover.[24] Further, within Lancaster Sound, space that's been opened up by increased rates of breakup of first-year ice could be responsible for an increase in the seasonal coverage of multi-year ice since 1971.[25] Overall, then, the pattern over more than 30 years is one of real but *uneven* reduction of sea-ice cover from subregion to subregion throughout the Archipelago. What about the "when" of a decrease in ice or no ice at all?

During the winter, solar radiation is absent and otherwise greatly reduced throughout the circumpolar North. Sea ice therefore increases in extent and thickness during the annual polar night and the twilight on either side of it. Global warming is a formidable countervailing force, but it cannot straighten the planet on its axis. It is true that it could eventually override the perennial phenomenon of freezing in the winter darkness. Arctic residents might therefore find themselves sweating in the dark on a December midday when power and air conditioners go out, but not anytime soon. So, the "when" of reduced sea ice comes down to the summer months.

The Canadian Ice Service defines "summer months" as the 17-week period between June 25 and October 15.[26] Obviously inexact, this is nevertheless a useful notion. The same may be said of the "shipping season". This term stands for the number of weeks during the 17-week summer-months period in which the sea-ice extent is a specified percentage of the total accumulated ice coverage.[27] For the western Arctic, the season is open when the ice extent is less than 60 percent of the entire ocean area within the region; for the eastern Arctic, the season is on when less than 30 percent is ice-covered; and for Hudson Bay, the figure is five percent.[28] The numbers vary because they are set with an eye to historic patterns of shipping under different circumstances: for example, the relatively high number for the western region reflects the fact that M'Clintock Channel and other areas clear infrequently, if at all. A

Northwest Passage that's open to shipping in summer is therefore likely to present *varied*, and not uniform, ice conditions.

Viewed in their totality for the period between 1969 and 2001, Canada's Arctic waters have experienced a two-week increase, from five to seven weeks, in the length of the shipping season.[29] If present processes were to persist for another three decades, we could be looking at an eight-week season nine years out of 10 in the 2030s.[30] This means that after 60 years of climate change and global warming, the shipping season in the Northwest Passage would have increased from five to eight weeks and possibly a bit more. Similarly, the ratio of shipping season to unfavourable weeks of the year would have risen from 5:47 to 8:44. Looked at closely, and with human rather than geological time in mind, the opportunity for navigation of the Passage by non-ice-strengthened ships is increasing steadily but slowly, not swiftly and dramatically.

Meanwhile, substantial and extreme interannual variation in sea-ice cover is also reported by the Ice Service.[31] According to the Service, one year's conditions may be twice as hard, or easy, as the next. There is no reason to expect that this kind of variability will abate.[32] This means that the shipping company that refuses the expense of ice strengthening will not be able to count on using the Northwest Passage from one year to the next. By the same token, the firm that does want to make regular commercial use of Arctic North American waters for summer-months intercontinental navigation will have to accept the need for ice strengthening even as the ice thins.

If all of this were not enough to call into question the vision of a readily navigable Passage, there is also the fact that the season may open and close at different times during the 17-week or 119-day summer. This kind of thing happens as wind and other forces move ice about, thereby pushing ice cover above and below the regional percentage limit. Accordingly, a seven-week shipping season today does not necessarily mean seven consecutive weeks, or 49 consecutive days, during which unstrengthened ships can expect to cross the Passage without difficulty. Instead, the window of opportunity may swing this way and that several times, and in different regions, in the course of the summer. This means that a commercial vessel wishing to sail the Passage on short notice will be at the mercy of ice conditions that could change at the last moment. The same applies to short-notice naval use.[33]

Unpredictability, not conditions clearly favourable to navigation, is the net effect of climate change on the Northwest Passage thus far.

Economics of Intercontinental Shipping

Ships come in many types and sizes. To simplify, let us set aside bulk carriers and tankers together with large and small "tramp" vessels that sail on no set route. Instead, let us focus on container ships such as those that use the Suez Canal or Panama on regular voyages between the Atlantic and Pacific. If climate change and thinning ice are to make the Northwest Passage into a viable alternative to existing intercontinental navigation, box ships will have to take up the opportunity. To simplify further, let us imagine a single vessel that is not ice-strengthened, 35,000 tonnes, and therefore relatively small (carrying 2,400 TEU or twenty-foot long containers). She was launched for about $35 million a few years ago, and sails for about $27,000 per day on a timecharter basis plus fuel. These costs do not include additional insurance and other Arctic-specific expenses. The timecharter rate covers amortization, regular insurance, crew, victualling, port, periodic drydocking, and other items, including a 12 percent return on capital. In addition, fuel is $170 per tonne and 40 tonnes per day, and there are Arctic-specific costs that will be further examined in this paper. Elsewhere I have called this ship the *Suzie Q*.[34] Unlike the tramp carrier fleet which travels the oceans from opportunity to opportunity, the *Suzie Q* is part of a highly competitive industry in which every day's sailing, gained or lost on a set and, therefore, economical route, is what it is all about. Sailing normally takes place 24 hours per day at a rate of 16 knots. Given what we know about sea-ice conditions in Arctic North America, how might the owners and operators of the *Suzie Q* evaluate the option of using the Northwest Passage today and in 2034? To be still more specific, how might the Passage fare, relative to Suez, for voyages by the *Suzie Q* between Yokohama and Rotterdam?

Sailing via the Passage saves roughly 3,600 nautical miles (nm) (nautical miles, distances and other numbers are approximate) over Suez. Given a constant speed of 16 knots and a daily operating cost of $27,000, the time saved over Suez, which is 9.5 days, would yield $247,000. In addition, there is a toll for the Canal (Canada does not exact a levy). For a laden ship the size of ours, the toll would come to some $125,000. Add time and toll saved, and use of the Passage would yield something like $372,000 per voyage over Suez for a *Suzie Q* moving at a steady 16 knots, equivalent to about 13 days' sailing. If this was the only issue, we would have to ask why the Northwest Passage is not already choked with summer-months intercontinental shipping when the going looks good.

A traverse of the Archipelago and its eastern and western approaches at a steady 16 knots could conceivably be achieved on occasion. But such a voyage would be so rare and so risky that in truth it is hardly worth consideration except to illustrate where the vision of an ice-free Northwest Passage takes us. Still, let us carry on.

First of all, a 16-knot voyage would be a last-minute undertaking, not a matter of regular routing. The decision to go north would depend on an ice-conditions forecast that was as favourable as it was reliable. It would have to be good not only for the duration of a transit of Arctic waters, but also for the voyage from, say, Yokohama to the Beaufort Sea.

Second, such a voyage would be rare because even with the requisite forecast in hand, the owner of an unstrengthened ship would have to accept the risk of an unexpected deterioration of ice conditions, fog and other causes for delay, accident, and a need to turn back. Major firms are not in the Passage on a regular basis because owners are not fools.

Third, ice conditions being what they are, 16 knots as a regular expectation is not reasonable. The *Suzie Q*'s owner would have to consider a reduced rate of progress through the Arctic portion of a total voyage of 7,600 nm. Remember, we are talking about an unstrengthened vessel. Suppose, therefore, that multi-year ice were to appear along the route and this, together with other factors, reduced the *Suzie Q* to an average speed of seven knots through some 2,000 nm of Arctic North America. In this case (168 nm per day) use of the Northwest Passage would require not five days at 16 knots, but nearly 12 days. At an operating cost of $27,000 per day, seven days lost over a dash through the Passage reduces the theoretical advantage over Suez to about $183,000, and that is when the trip can safely be made. The point is that the less the *Suzie Q* is assured of a rapid transit of Arctic North America, the better Suez – and Panama – looks.

Fourth, there are additional expenses associated with summer-months Arctic navigation. Some have to do with the services of an ice navigator and the purchase of satellite-based ice reconnaissance. But the outstanding factor being considered here is the cost of insurance against the following: accident, holing, getting stoved in, environmental damage, and other liabilities. Insurers are said to charge two to three times the open-water rate for ships using the Northern Sea Route.[35] It is hard to believe the rate would be less for last-minute transits of the Northwest Passage –

particularly when the bathymetry or charting of Canadian Arctic waters is far from complete, some of it dating back to the search for the Franklin expedition. Furthermore, it has been reported that insurers already determine the length of the shipping season in Hudson Bay. If so, they will certainly have something to say about the high Arctic season, when it is open, and when, if ever, the *Suzie Q*'s owner might seek a rapid transit. Since there is no established market for the kind of coverage (including the size of the deductibles) being considered, it is hard to say what the effect of Arctic-specific expenses might be in comparison. Let me therefore guess that it lops a further $100,000 off the advantage to Suez, bringing the gain to some $83,000. For this amount of money, reputable firms are not going to risk high-value ships on a once-a-year Arctic voyage.

Fifth, ice strengthening is the way around the difficulties presented by the Northwest Passage, but it is expensive. To bring the *Suzie Q* to Baltic or St. Lawrence capability, which is merely to break first-year ice, would add amortization costs that wipe out the advantage of avoiding Suez. To be sure, it would increase the safety and the reliability of one summer-month transit of the Passage when sailing conditions are ideal, which is to say when there is assuredly no multi-year ice along the route. For the rest of the year, however, amortization costs of a $5 million or greater upgrade would have to be borne in open-water sailing and navigation in first-year ice conditions outside the Arctic. The *Suzie Q*'s owner also would not be able to rely upon a round trip through the Archipelago in a single season that normally runs from 49 to 56 days. Ice-condition uncertainty and turn-around time in port make this improbable.

Nor is the solution for the owner to add further strength by putting the likes of the *Suzie Q* aside and acquiring purpose-built, Arctic-class ships. Such vessels would be able to pull back, charge the ice repeatedly, and make way in year-round polar navigation. Why go to the even-greater expense of Arctic class when Suez and Panama offer cheaper alternatives for year-round commercial shipping between the Atlantic and Pacific? Security and geopolitical reasons could be advanced. If they turned out to be persuasive, we could see an increase in the volume of intercontinental traffic through the Passage by major firms. But the increase would owe little, and more likely nothing, to calculations about thinning ice now and over the next three decades. On the contrary, the idea would be to override not only ice-related uncertainties but a good many market variables in intercontinental shipping.

Directions

Where might all of this take us? Let us provisionally separate the commercial from the security implications, and then consider them together for Canada (more commercial than security-related) and the United States (more related to security than commercial).

On the commercial side, predictions of an ice-free Northwest Passage and heavy use of Arctic North American waters by major shipping firms in the 2020s and 2030s cannot be sustained. Instead, Canada would seem to face a gradual and irregular increase in the volume of occasional or opportunistic shipping. The bulk of these vessels would move in and out of Canadian Arctic waters without transiting the Passage, for example in moving between Europe and the port of Churchill in Hudson Bay. As well, we may anticipate the appearance of fishing vessels in greater numbers, and growth in the bulk transportation of Arctic hard minerals, for instance southwards from Baffin Island. All the while, regular summer resupply of Arctic communities from the south will continue, as will cabotage. Some growth in intercontinental navigation is also to be expected as the following occur: barges and drilling rigs are towed between the Pacific and Atlantic oceans, tourist vessels appear in increasing numbers, marine scientific activity unfolds, yachts and adventurers make their way through the Archipelago, and tramps appear. But as long as unstrengthened container ships sail on set routes, variable ice conditions will continue to deny the *Suzie Q* or her bigger sisters the opportunity to make the Northwest Passage a regular route.

In sum, unless ice conditions improve more rapidly in the Canadian Archipelago than in the Arctic Ocean, which seems unlikely, the risks and costs of using the Northwest Passage for summertime intercontinental voyages by non-strengthened ships will outweigh the advantages as long as other, safer ways are available. By the time the ice is reliably easier in the Passage, conditions in the Arctic Ocean, if not also in the Northern Sea Route, promise cheaper and safer transpolar summertime sailing. This will be especially so as long as Ottawa continues to shun icebreaker-supported convoys and, therefore, to transfer large added costs of Arctic operations to shipping firms by insisting that vessels conform to a regime of independent navigation. All the while, use of the term "ice-free" will be the enemy of clear thinking about Arctic waters. It should be banned outright.

Instead, the likely scenario for major new shipping activity – a scenario that is independent of climate change and thinning ice – is for north-south voyages carrying Arctic hydrocarbons to southern markets, principally LNG to the east coast. Any such ships are certain to be massive, ice-breaking, and readily able to operate year-round. Should they appear, it would be for security as well as commercial reasons. Given that the United States and Canada may wish to move LNG eastward through the Passage, the two countries could encounter difficulty and conflict in the years ahead. Such a prospect will persist as long as the two governments fail to preempt foreseeable trouble by enlarging their Arctic marine cooperation.

Meanwhile, Canada has maritime security interests of its own that cannot be met without wider collaboration with the United States. One of these arises from the occasional shipping that has been mentioned. Though other vessels, such as fishing boats, could have persons aboard with nefarious purposes, experience suggests that tramps are more likely to be connected to human smuggling (stowaways included), drug trafficking, illicit arms sales, and the like. There is also the potential for terrorist use, including the movement of weapons of mass destruction or merely surface-to-air Stinger missiles for use against commercial aircraft to the south. A natural reaction to the thought of terrorist use of Arctic waters is to scoff. Nevertheless, Canada is making a real effort to secure its southern border and its share of the eastern and western coastal approaches to North America. Failing to adequately cover its Arctic waters as well is to maintain a path of least resistance for terrorists and their associates.

Whereas Canada had marine safety regulations that make it mandatory for mariners to provide a minimum of 96 hours' advance notice and crew and passenger lists prior to entry into Canadian waters, no such measures were applied in the Arctic offshore. The reason was simple: Because officials were concerned that Canadian insistence on mandatory notification could be interpreted by the United States as an assertion of Canadian jurisdiction and therefore sovereignty over the Northwest Passage, Ottawa chose not to act for security when it came to the Arctic marine environment. Now, following consultations with the U.S. Coast Guard, Canada has begun to act by implementing the International Maritime Organization's International Ship and Port Facility (ISPS) Code around and in its Arctic waters. [36] In my view this should only be the beginning of a new series of moves for enhanced Canada-U.S. cooperation in the Arctic waters of North America.

As for the United States, whatever its current interests in commercial use of Arctic waters, naval uses are relatively amenable to long-term planning, which means something coherent can be done about them now. This paper will touch briefly on the naval dimension, but first will include a reminder that commercial and security considerations travel together. Middle East instability, the potential for alternative hydrocarbon reserves to appreciate in strategic value, insecurity of Suez, and a gradual increase in the accessibility of Arctic waters for commercial and naval use both *from* sites in Arctic North America and *through* the Northwest Passage – much of this is a familiar geopolitical agenda. But not all, and the familiar itself is becoming less predictable. In circumstances such as these, the United States, particularly the U.S. Navy, would seem to have good reason to look ahead and begin addressing priority items that take significant time to complete.

This is certainly not the place to consider capabilities, for example, whether one or more ice-strengthened frigates or other icebreaking warships should be acquired by the United States and, if so, what the level of ice strengthening should be. Or whether additional icebreakers should be procured to support extended operations by non-ice-strengthened, battlegroup-sized forces, and so on. Rather, the critical early-days problem of U.S. Arctic naval development as a surface, air, space and subsurface power is to arrive at a proper understanding of the emerging situation in the region and what it demands. In order for the United States to accomplish more in the Arctic theatre than it can do on its own, it is essential for Washington to coordinate with regional allies and friends, as well as come to a realistic consensus on ice conditions over the next few decades. Coordination with other nations could address diverse matters such as threat assessment, interoperability, joint exercises and logistical support, Arctic science, bathymetry, and pre-emptive management of potential conflicts. On threat assessment alone, are we on balance faced with conventional naval missions such as keeping sea lanes of communication open and escorting commercial convoys in projecting and otherwise exercising naval power in the 21st century Arctic? Or is the need one that calls primarily for reliance on intelligence and surveillance in detecting and countering non-traditional threats? An assessment of this kind will have to begin somewhere; why not have it begin with discussions between Canada and the United States?

Without necessarily aiming at anything like a maritime NORAD, Ottawa and Washington could start to broaden their Arctic commercial and security cooperation on many fronts: prospects for marine transportation of hydrocarbons; joint Arctic naval and combined-forces exercises, as ice conditions permit; border security; joint assessments of the Arctic commercial and naval environment to the 2040s; potential for coordination and division of labour in likely naval roles and missions (e.g. minesweeping, search and rescue); outlook for forward basing and logistical support; consensual knowledge of the interface between naval operations and environmental protection; role of indigenous peoples and local economic impacts of naval- and shipping-related development; and conflict prevention. As long as the Northwest Passage remains a potential flashpoint between Canada and the United States, Arctic commercial and security cooperation between the two countries will be wanting.

Ottawa could take the initiative and propose new Arctic marine cooperations talks with the United States. Though others will surely have plenty to add, I would suggest three main points to begin with. First, Ottawa should propose an extension of the 1988 Canada-U.S. icebreaker cooperation agreement[37] to include commercial-vessel and warship navigation in the Arctic waters of North America. The 1988 accord is without prejudice to the position of either country in international law. It works very well in practice. To extend it to commercial vessels and warships would not change anything in law. Neither country would alter its legal position, much less surrender its judgment on the contending claims. Instead, the two would build on existing practices to defuse renewed conflict over the Northwest Passage before it arose and do so without effect on the fundamental legal issues if ever they came to court.

Second, in a related act of conflict preemption, Ottawa could encourage Washington to take a homeland-security look at the practical consequences of the U.S. legal stance on North America's Arctic archipelagic waters. That stance would have us run an international strait through the continent's northern perimeter. Whatever the security impact of an enraged and bitter Canada following a "victory" for the United States in this matter before 9/11, is it not the case that the two countries have collaborated more closely in matters of security since that date? Do they not have an ever-greater need to avoid conflict? I say they do. Canada should therefore act to close the perimeter in the Archipelago and its approaches in conjunction with the United States. If necessary, Canadian activity in rule-making and enforcement within and near the

Archipelago could again be affirmed without prejudice to U.S. and Canadian claims on the status of the Passage in international law.

Finally, naval coordination along lines already considered could form a third basket in Canada-U.S. talks on Arctic security and commercial cooperation.

Some Canadians and some Americans are certain to oppose any broadening of without-prejudice collaboration in Arctic waters. Ironically, the opposition may well include those on both sides who join in favoring an overwrought view of climate-change effects on archipelagic sea ice. It is high time to take the field from the exaggerators. Let us come to a sensible understanding of what lies ahead, and of what we can do together to make the most of it.

Notes

[1]Office of Naval Research, Naval Ice Center, Oceanographer of the Navy, and the Arctic Research Commission, "Naval Operations in an Ice-free Arctic. Symposium 17-18 April 2001" (Arlington VA: Office of Naval Research, 2001). Also of special interest is Usha Lee McFarling, "Melting Ice, Winds of Change," a two-page spread in the *Los Angeles Times*, 19 January 2003 and, briefly, Andrew C. Revkin, "Unfrozen North May Face a Navy Blue Future," *The New York Times*, January 13, 2004. Both these news stories cite the ONR report, with the latter story also reporting that many scientists regard the Navy's concerns as legitimate.

[2]Rob Huebert, "Climate Change and Canadian Sovereignty in the Northwest Passage," *Isuma* 2 (Winter 2001-2002), 86-94, and "The Shipping News Part II: How Canada's Arctic Sovereignty Is on Thinning Ice," *International Journal* 58:3 (Summer 2003), 295-308. The second of these articles was in response to a piece of mine. In it Huebert lets up a bit but does not surrender the expectation of substantial international shipping through the Northwest Passage. My paper was "The Shipping News: Canada's Arctic Sovereignty Not on Thinning Ice," *Ibid.* 58:2 (Spring 2003), 200-222.

[3]Huebert, "Climate Change and Canadian Sovereignty," 93.

[4]Quoted in McFarling, "Melting Ice."

[5]For some recent examples, see "On Thin Ice," *This Magazine* 37:3 (November/December 2003), 32-37; "Bearings: What's Happening to Our Arctic Sea Ice," *Nature Canada* 32:3 (Winter 2003-2004), 28-29, which speaks of a "year-round passable Northwest Passage"; and Editorial, "Guarding the Arctic," *The Toronto Star*, April 5, 2004.

[6]"Naval Operations," 3 and 7.

[7]*Ibid.*, in Appendix A, 8.

[8]*Ibid.*, main text, 12.

[9]*Ibid.*, 11.

[10]*Ibid.*, 3.

[11]*Ibid.*, 43.

[12]*Ibid.*, 44.

[13]Huebert, "Climate Change," 91.

[14]*Ibid.*

[15]Huebert, "Shipping News Part II," 298.

[16]*Ibid.*, 303-304.

[17]Huebert, "Climate Change," 91. See also "Shipping News Part II," 300-301.

[18]"Climate Change," 94.

[19]*Ibid.*, 91.

[20]John Falkingham, Richard Chagnon,, and Steve McCourt, "Trends in Sea Ice in the Canadian Arctic," paper presented at the 16th IAHR International Symposium on Ice, Dunedin, New Zealand, December 2-3, 2002, 1.

[21]*Ibid.*, 4.

[22]John Falkingham, Canadian Ice Service, communication to the author, January 27, 2003.

[23]"Trends", 4-5.

[24]*Ibid.*, 6.

[25]*Ibid.*, 6-7. Sea also John Falkingham, Richard Chagnon, and Steve McCourt, "Sea Ice in the Canadian Arctic in the 21st Century," paper presented at the 16th International Conference on Port and Ocean Engineering under Arctic Conditions, Ottawa, August 12-17, 2001, 8.

[26]"Sea Ice," 5.

[27]*Ibid.*, 6.

[28]*Ibid.*, 6-7.

[29]Falkingham, communication to the author, January 23, 2003. See also "Sea Ice," 6 (figure 6).

[30]*Ibid.*

[31]"Trends," 2, and "Sea Ice," 3, 6, and 8.

[32]"Trends," 2, and "Sea Ice," 3.

[33]The ONR report ends with mini-scenarios for new U.S. naval operations in Arctic waters. One of these envisages the dispatch of a carrier-based task force through the Northwest Passage in the 2020s. The aim would be rapid reinforcement from Norfolk to the China Sea in the event of a Taiwan-China confrontation. Among the preconditions for a successful surface naval transit by very high-value unstrengthened ships are extraordinarily favourable, which is to say improbable, ice and weather conditions. Indeed, this is what the scenario supposes: "The NW passage is ice-free in the summer and commercial shipping routinely transits." "Naval Operations," Appendix B, 1.

[34]This in a piece currently being considered for publication: "Pathetic Fallacy: That Canada's Arctic Sovereignty Is on Thinning Ice."

[35]Julius Strauss, "Polar Thaw Opens Arctic Sea Route," *The Daily Telegraph*, March 4, 2004.

[36]Transport Canada, "Security Requirements Strengthened for Vessels, Marine Facilities and Ports," *News Release* H042/04, May 21, 2004.

[37]Agreement Between the Government of the United States of America and the Government of Canada on Arctic Cooperation, 11 January 1988, *International Legal Materials* 28 (January 1989), 141-142.

Coming to the Arctic: Oil, Ships and UNCLOS Plus Risk and Research

George B. Newton[*]

In the first four years of this new millennium, the public has heard numerous predictions about the Arctic Ocean. The predictions generally focus around one thing: At some point during the 21st century the Arctic will be covered with so little ice that commercial vessels will able to use the Northern Sea Route, north of the Russian Littoral, or the Northwest Passage, through the Canadian Archipelago, for international commerce, intercontinental commerce or both. The degree to which the ice cover will decrease, and the rate at which it will occur, is the subject of active scientific debate. But there is solid consensus on the fact that change (warming) in the Arctic is occurring.

This fact naturally stimulates many questions such as the following:

- Does improved access to the Arctic Ocean lead to commercial exploitation?
- What are the commercial advantages of a polar shipping route?
- Are there other circumstances to force this activity?
- Are there hazards and concerns about which we should be concerned?

A more accessible Arctic Ocean appears quite logical, practical, and beneficial for a number of reasons:

- Because nine out of 10 people reside on continents that border the Arctic Ocean, the world has a decidedly Northern Hemisphere "focus."
- Preliminary predictions from the internationally based Arctic Climate Impact Assessment (ACIA) indicate that usable polar routes will be available later in this century. (See figures 1 and 2.)

[*] Chair, U.S. Arctic Research Commission.

- *Lloyd's Register of Shipping* currently estimates that there are over 89,000 vessels in the world weighing over 100 metric tons (MT), which makes them sea-worthy in the Arctic Ocean.
- Sea going interests will look to the polar routes as a way to reduce costs, because transit time and distance savings of nearly 40 percent can be anticipated (as shown in figure 3).

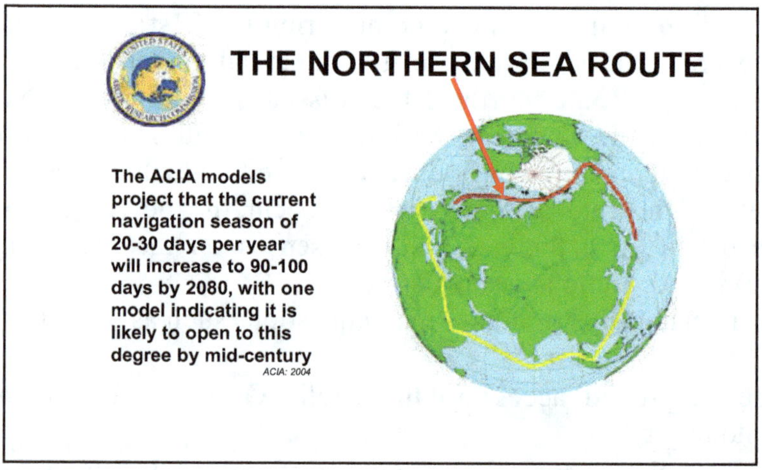

Figure 1

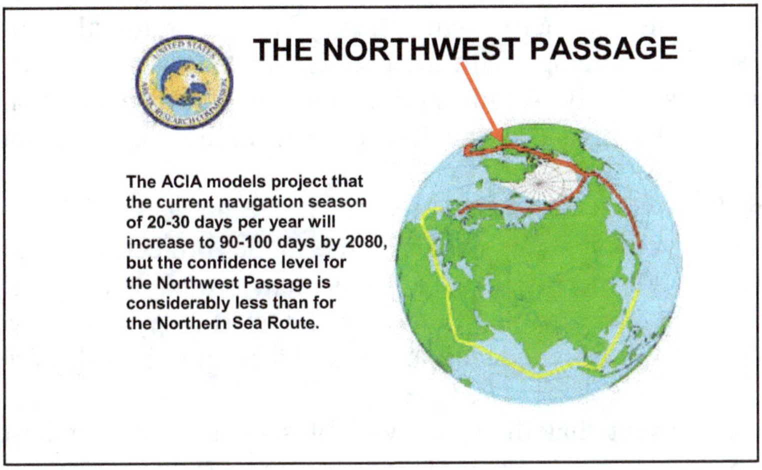

Figure 2

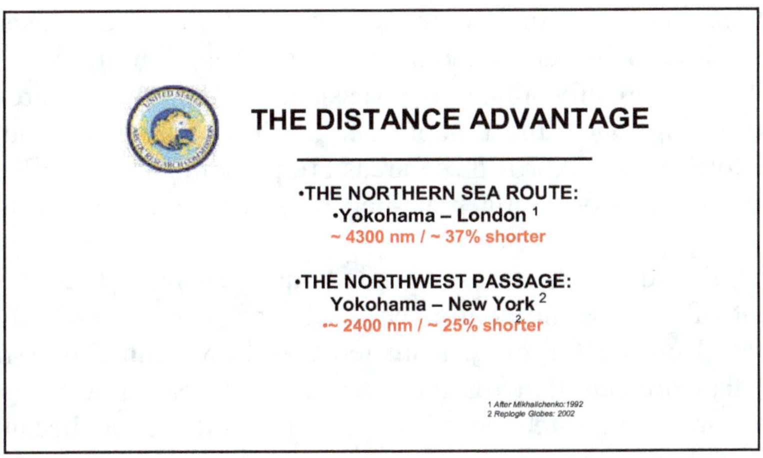

Figure 3

The ACIA predictions can be viewed as making good sense when one examines the statistics for the last 100 years of the seasonal trend in sea ice extent for the Northern Hemisphere. The trends for all seasons are toward less areal coverage with the greatest decrease shown in the Arctic summer months — July, August and September. The data shown in Figure 4 was one of the inputs to the five climate models used in the ACIA report.

Figure 5 provides projections on Arctic Ocean accessibility for ships during the 21st century. For example, the model projections suggest that by the year 2100, there will be a 25 percent probability of unimpeded access along the Northern Sea Route for almost 180 days of the year. Use a probability of 75 percent accessibility and the number of days decreases to 90.

Despite the fact that there are variables, it is clear that based upon what has been observed in the past half-century or more, warming in the Arctic is occurring. It also can be assumed that there will be sufficient catalysts in the environment to ensure that this trend continues. This trend may slow somewhat, but every scientifically based suggestion indicates that temperatures will continue to rise throughout the 21st century. It is my opinion, therefore, that more and more of the Arctic Ocean will become ice-free each year, and will be that way for longer periods of time. This will occur mainly along the Russian Littoral, in and around the Canadian Archipelagos, the Lincoln Sea north of Greenland, and off the Alaskan North Slope. All of these areas are predominately shallow water areas (≤200 meters) or continental shelf areas where annual ice normally forms.

In fact, the Arctic Ocean as a whole is quite shallow. It contains over 25 percent of the oceanic continental shelf area that exists in the entire world. Based on the forgoing, I am led to follow a timeline (shown in figure 6) that projects that routine internationally based Arctic shipping will occur in about three decades. Shipping will ensue because it is feasible, safe and economically attractive.

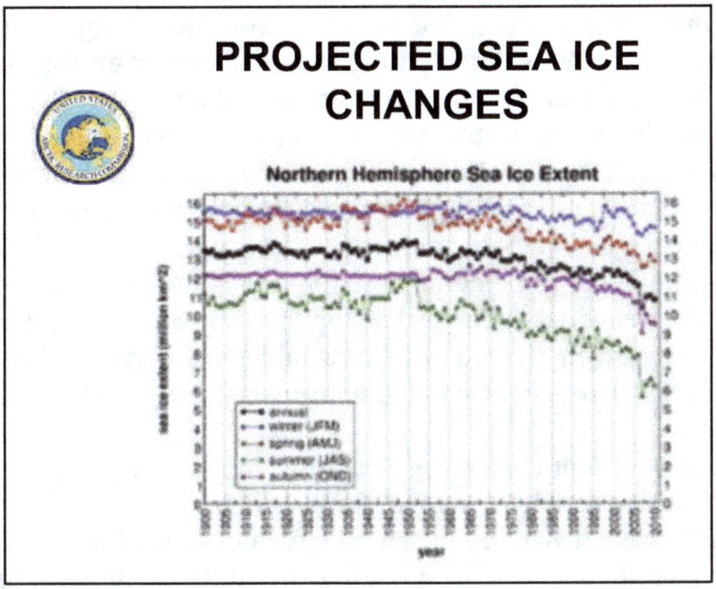

Figure 4

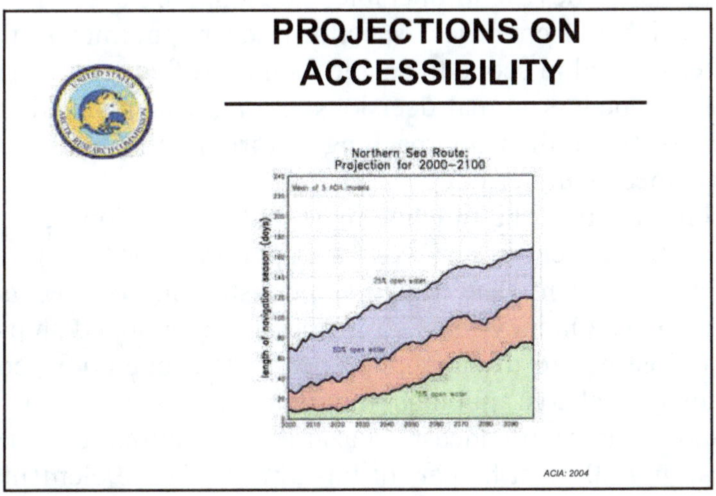

Figure 5

The controlling factors that must be considered for all this to be realized are: (1) the considerable time required to prepare for this level of shipping, and (2) making the decision to seriously exploit the Arctic transportation routes. For the first factor, one need only consider the Arctic commercial fleet. A fleet of ships large enough to be economically feasible and also safe enough to survive the dynamic and unpredictable Arctic will take a considerable amount of time to design, build, and launch in sufficient numbers to be commercially viable and to meet demand. For example, experience has shown that for the first of a new class of ships (which new, ice strengthened vessels would most certainly be), the U.S. Navy requires 10 to 15 years to go from concept development to full operation.

Next, consider that the average life of a ship is 30 to 35 years, and one realizes that these newly designed ships will be, if projections are even roughly accurate, in service when the Arctic is well into becoming a usable transportation route. Even if someone starts the design phase today, there will be a rush to assemble an adequate fleet within a timeframe that would allow the timely exploitation of these routes.

Having addressed ships, next consider all the elements that must be brought together to devise an operable, safe shipping system. In order to effectively satisfy all countries' specific trade requirements, the system must be international in scope to enable the level of cooperation necessary to cultivate acceptable mutual decisions and agreements. This process is much more difficult than approaching a fork in the road, and simply deciding whether to turn "right" or "left."

To accomplish this, the timeline must include developing and gaining approval for the concept of an Arctic transportation system. But approval in itself is not sufficient. The design and construction of the system must also be funded. The money must cover the aforementioned ships, plus the necessary infrastructure to support proper system operation, control and safety. Buoyage, charts, pilotage requirements, escorts, contingencies, warning and weather broadcasts, hazard information, and search and rescue capability all must be part of this system. These elements must be accomplished under the regulatory umbrella of the International Maritime Organization (IMO) Polar Code and of the Law of the Sea, which permit nations to establish rules (mainly addressing ship construction and ocean pollution) for passage in the territorial seas or exclusive economic zones (EEZs) of coastal States. Such an international system, even if it is not accomplished through the efforts of many nations, must at least be understood and accepted by them.

That brings me to the most important, yet up to now most under-appreciated factor in the system . . . Oil!

With easier access to the Arctic Ocean that enables increased commercial shipping and allows Arctic rim nations to consider on- and off-shore exploitation of the abundant fossil fuel resources (mainly oil) (see figure 7), the concern for an oil spill in high-latitude, ice-infested waters becomes very real (as shown in figure 8). It is an event the world is ill prepared to face.

From statistics found on the U.S. Coast Guard Web site it is clear that throughout the temperate world, in spite of improved technology, tighter regulations, and heavier penalties, significant oil spills continue to occur.

Initially looking at worldwide numbers since 1960, oil spills of at least 10,000 gallons (roughly 34 metric tons or 227 barrels of oil) have occurred off the coasts of 112 countries (high seas or territorial waters). Counting only those that occurred in what are best described as high traffic areas (Florida Straits, English Channel, Straits of Malacca, Mediterranean Sea, etc.), there have been 982 spills of at least 10,000 gallons. That works out to 28 per year or 2.3 a month. (See figure 9.)

Similarly sized oil spills in the United States, occurring on both land and sea during the 1986 to 1999 period, numbered 370 or just over two per month. (See figure 10.)

Thus, even though the Law of the Sea gives coastal States the authority to set regulations and controls to prevent pollution for vessels embarked purely on innocent passage (Article 21), transit passage (Article 42), or in ice covered waters (Article 234) (as shown in figure 11), regrettably, one must assume that based on statistics a significant oil spill will still occur some day in the Arctic Ocean sea ice.

Unless all Arctic rim or Arctic sea-going nations act positively, it will be an extremely damaging event, both practically and politically. Why? Because the world has very poor understanding and skill when it comes to the best way to control and mitigate an oil spill. Spill response capability is limited, distant, and slow. Much research is required to improve all aspects of countering an ice-related oil spill. It is important to note that already the United States is conducting oil drilling and recovery operations in the Arctic offshore — in the Cook Inlet near Anchorage, Alaska, and at the North Star Production facility off the North Slope.

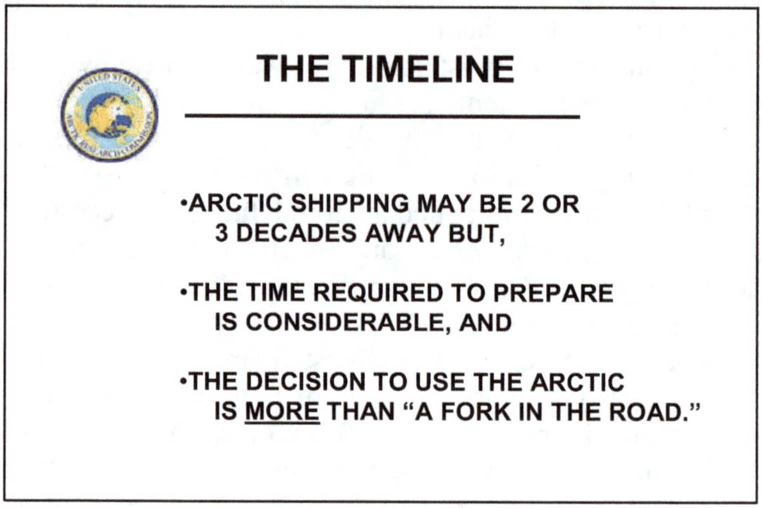

Figure 6

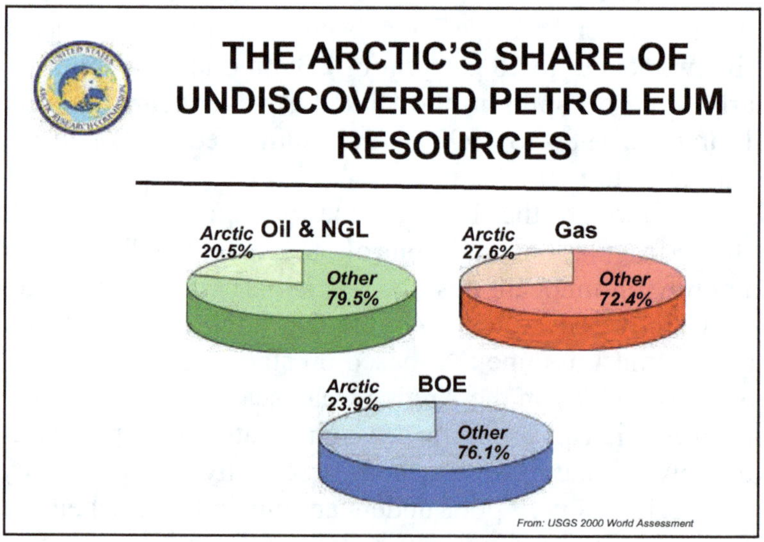

Figure 7

Figure 8

OIL SPILL FREQUENCY
(WORLDWIDE)

- SPILLS OF AT LEAST 10,000 GALLONS (34 METRIC TONS / 227 BARRELS) HAVE OCCURRED OFF COASTS OF 112 COUNTRIES SINCE 1960. [1]
- SPILLS ARE CONCENTRATED IN HIGH TRAFFIC AREAS.
- BETWEEN 1960 – 1999: 982 SPILLS. [2]

AVG: 28 / YEAR

1 Aspen Publishers: 1999
2 Etkin: 1997

Figure 9

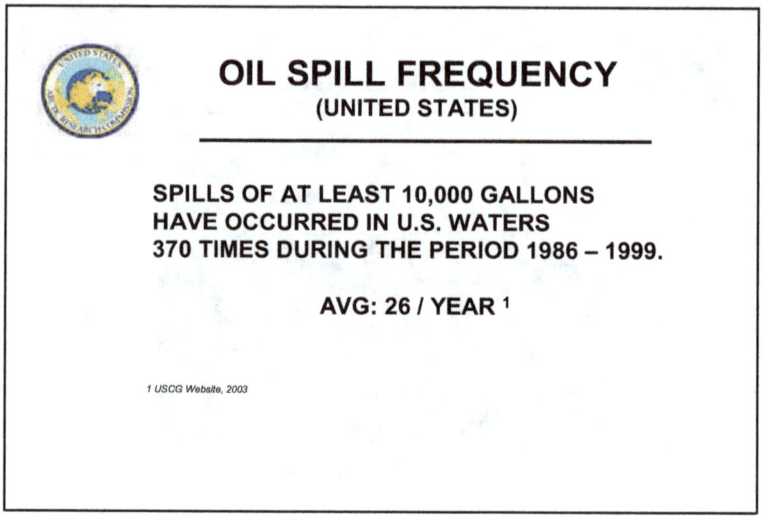

Figure 10

Figure 11

One must also be sensitive to another important Arctic fact. Oil spill response operations in ice and open water are fundamentally different. The most notable differences are usually challenges created by the cold. And when assessing spills in ice, the response timeline becomes much more critical:

- Immediate response is essential.
- The regulations that control and direct spill response must be more flexible, allowing greater on-site decision-making latitude.
- Education of, and outreach to, the populous is very important, stressing possible impacts and predicted responses.
- Involved organizations must know in advance the effects and uses of biological sciences to mitigate the net environmental effects of a spill.
- Perhaps most importantly, research must be conducted in the field – in the cold and in the ice.

The foregoing concerns and the attendant, specific research needs they engender have been recently documented in a report entitled *Advancing Oil Spill Response in Ice-Covered Waters*, published in March 2004, and jointly sponsored by the U.S. Arctic Research Commission and the Prince William Sound Oil Spill Recovery Institute. The report, which can be viewed on the Commission's Web site (www.arctic.gov), was the first U.S. effort to identify the unique research requirements for improving current spill response capability in ice. The baseline for the research plan was derived from a January 2000 International Workshop on oil in ice, held in Anchorage, in which the many scenarios involving this potentially disastrous occurrence were addressed. Peers in government, military, academia and industry from five Arctic nations reviewed the USARC/OSRI report.

Oil spilled on or in ice is complicated, not just because of the ice itself, but because of the many different fates the spill may realize in the ice environment; almost every one of which requires a different approach for clean up. The following are several different, unique pathways in which spilled oil may mingle with ice:

- Oil may pool on melt water ponds in the summer,
- Oil can migrate up into brine channels in sea ice,

- Oil can become encapsulated in ice during winter,
- Oil can become trapped in free-floating ice rubble,
- Oil may fill a lead,
- Oil can be trapped and pool under the ice,
- Oil can pool on top of the ice (and under the snow), and
- Oil can be absorbed by snow.

Combining these outcomes of oil and ice interaction with the fact that ice movement usually follows the ocean currents, it is clearly challenging to predict where the bulk of the oil is destined. Yet, this seemingly simple, but unsolved problem continues to confound scientists and makes clear the need for continued research.

Relevant and varied research requirements include the following:

- chemical herders and in-situ burning;
- mechanical systems enhancement;
- dispersants;
- detection of oil in ice by remote sensing;
- oil deflection;
- transfer of oil and oily waste; and
- simulants and field testing.

Chemical Herders and In-Situ Burning

This combination of methods for spill response is complementary. New chemical herders would thicken the oil to allow in-situ burning to be more effective. Of all oil-in-ice responses, in-situ burning is without question the most effective. But it must be initiated very soon after the spill. Not only would improved chemical herders enhance the best response, it is also a goal for the herders to increase the operating window during which burning can achieve maximum effectiveness. Herders also hold promise to improve the effectiveness of several other response techniques.

It should be noted that there are currently strong objections from the environmental community and most governments about in-situ burning. That is why improvement in spill response requires pre-approved methods (defined by government regulation or law).

Enhancement of Mechanical Recovery Systems

Mechanical recovery systems are most commonly used for oil spills in temperate, ice-free waters. They work poorly, or not at all, when ice is present. Narrow swath width and speed of advance limit their use in ice. Research is needed to expand their operating window when ice is involved and to combine their use with other spill response techniques that show greater promise, such as herding and oil deflection. Unfortunately, at present there is low confidence for progress in this area.

Dispersants

Dispersants are an important response technique in open water. However, no effective dispersant exists for viscous or cold oils. In addition, dispersants are most effective when subjected to high mixing energy, which is more difficult to achieve in ice-infested water. Stratification of water around the mouths of rivers creates a need for a dispersant or multiple dispersants that are effective in different water densities. There is hope that a cold adapted dispersant combined with energy insertion by a large vessel, such as an icebreaker, will achieve improved performance in water of consistent density.

Detection of Oil in Ice by Remote Sensing

It is critical to know where the spilled oil is located in order to affect a proper response since, as was discussed earlier, there are many fates when there is a combination of oil and ice. The current inability to locate oil in ice quickly is a serious shortfall. Aerial systems do not show much promise, but synthetic aperture radar is one solution now being considered. Land- and water-based systems only perform marginally better. Acoustics, gas detection, infrared, and optical technologies (now used in open water spills) all are being studied for their adaptability. Confidence in future development is at best guarded, and yet spill detection is the very first step in spill response.

Oil Deflection

Methods that separate oil and ice on the water will enable increased encounter rates for possible recovery by mechanical systems or facilitate in-situ burning inside a fire boom. It is very frustrating that even low concentrations of ice seriously degrade the performance of most skimmer systems. Recent experiments using the propeller or thruster wash from a large ship showed only modest effectiveness.

Transfer of Oil and Oily Waste

Attempts to pump oil and oil-coated ice, even in small chunks or as slush in the cold, have had very little success. There is little baseline knowledge in this area, but use of emulsifiers is being considered. Because pumping oil is integral to the oil production process in all climates, there is considerable optimism that suitable, effective technologies can be developed in this area.

Simulants

The United States bars field testing for oil spill research even when small quantities are proposed (i.e., three barrels). The U.S. Environmental Protection Agency, in responding to several requests in the 1980s and 1990s, has consistently refused to allow field testing (for in-situ burning) on the North Slope of Alaska, even though there has been unanimous approval by local governments, the state, naturalists, and several involved U.S. government agencies. The most recent request was made in 1992.

In response, stakeholders have turned toward developing an environmentally friendly oil simulant to serve as an oil proxy. Items used in attempts to replicate oil have included oranges, popcorn, hula-hoops and canola (vegetable) oil. Unfortunately, nothing thus far has fully replicated properties of oil. The simulant must be non-toxic, non-sticky, environmentally friendly, and rapidly broken down.

Field Testing

Despite these efforts, a bona fide field-testing program remains necessary to ensure accurate research results in other research areas. The United States, however, is not the only nation that refuses to permit field testing for oil spill research. Yet such tests are needed to improve response procedures in ice for the following critical reasons:

- validation of laboratory experiment results;
- training of spill response personnel, just as fire fighters occasionally burn down structures under controlled conditions;
- development of stakeholder (public, industry, government) confidence in response methodologies;
- improvement in the understanding of the fate of oil in ice; and
- development of effective procedures for particular technologies.

As far as research for this paper would allow, only two deliberate, controlled oil spills in ice have ever been permitted: one in Norway in 1986 and one in Canada in 1993. They demonstrated clearly that of all the elements in the research program described here, field testing is absolutely essential.

Thus, one can now comprehend the challenges of combating oil spilled in ice. No matter how careful we are statistics tell us that there will be spills in the Arctic Ocean as access improves and as transportation and oil exploitation operations grow. International cooperative preparation is critical, particularly as the tenets of the Law of the Sea prescribe.

The United States, specifically, has no government response plan for spills in the Arctic Ocean. The country's primary responder, the Coast Guard, has its nearest location on Kodiak Island in South Central Alaska, nearly 800 air miles from the North Slope. While the nearest U.S. icebreaker's homeport is in Seattle, the ship itself could ideally be located in the Arctic or, conversely, in the Southern Ocean.

In summary, the needs are clear. Any Arctic transportation system plans must include oil spill response methodologies based on research and international cooperation.

CONCLUDING REMARKS

Concluding Remarks

John Norton Moore[*]

We have been privileged over the last several days to hear world-class experts on important issues relating to the Arctic and to international energy policy. We have not necessarily found the solutions to all problems, but we have certainly identified an interesting set of problems that need solutions. Some things do suggest a level of agreement in the conference – the great importance of international cooperation in solving these issues, the great importance of international law, the great importance of the special achievement represented by the Law of Sea Convention itself, and perhaps even more broadly, the great importance of the rule of law both internationally and nationally as we move forward to meet common goals for economic development and environmental protection.

Let me join Professor Skaridov in thanking the world-class participants, all of the panelists and those who have participated from the audience as well. I would also like to offer a special thanks to our Russian hosts. Professor Skaridov, be most assured this has been a magnificent surrounding for this conference, and all of us are grateful to you and our Russian hosts. I have already extended special thanks to you and Judge Anatoliy Kolodkin, but there are a number of other Russian hosts who participated in this conference who should be thanked as well. Among those is Alexander Etimal, one of the sponsors of this conference whom I had the privilege of meeting. We appreciate what he did in enabling this conference to take place. I would also like to thank Glendamere Ivanov, the hotel director, without whom we would have been sleeping in the streets. We greatly appreciate the housing achievement, which at this high season of the year is no small accomplishment; Tatiyana Mordvinova, the senior conference manager, Anatoli Butkoca, the Conference Manager, Nabila Notiava, the visa Assistant manager, Alina Ilyina, who assisted with translation, and others. Thanks to all of you for hosting this conference.

---

[*] Professor of Law and Director, Center for Oceans Law and Policy, University of Virginia School of Law.

www.ingramcontent.com/pod-product-compliance
Lightning Source LLC
LaVergne TN
LVHW010624300926
849162LV00001B/56
9789004144033